Flowforms

Flowforms

The Rhythmic Power of Water

A. John Wilkes

Floris Books

First published in 2003 by Floris Books
This second edition 2019

Frontispiece: The original sketch 2002 for a radial
Flowform complex, commissioned by Henry Nold of
Darmstadt as a celebration of the life of Viktor Schauberger.

British Library CIP Data available
ISBN 978-178250-589-1

Contents

Part 3: Applications and Research

Foreword

A biographical note

My first meeting with Goethean science and the work of Rudolf Steiner came through an introduction by Sir George Trevelyan, at that time Warden of Attingham Park Adult Education Centre, to Dr Ernst Lehrs, scientist, educator and author of *Man or Matter*. This was in June 1951. My acceptance, earlier the same year, by Professor Frank Dobson of the Sculpture School at the Royal College of Art brought me eventually to London, where I continued my studies for three years under Professor John Skeaping, Leon Underwood and others. During this period I was able to meet many personalities connected with a way of thinking which offered new insight into the spiritual background of the twentieth century.

One important meeting was with the mathematician George Adams who offered to introduce me to some aspects of modern projective or synthetic geometry, in contrast to the analytic geometry in which we were also being instructed at the College. During these years George Adams was developing a collaboration with Theodor Schwenk, a pioneer in water research who gained wide recognition with his book *Sensitive Chaos*, and on the basis of this they later founded the Institute for Flow Sciences in Herrischried, Germany, with a group of scientists and supporters.

After six years in freelance design and part-time teaching at Bromley College of Art, I was invited in 1961 to join the newly founded Institute as George Adams' assistant. With my practical skills and some knowledge of projective geometry, I was to make apparatus for research purposes. An aspect of George Adams' work was concerned with 'path-curve' surfaces described initially by Felix Klein and Sophus Lee in the nineteenth century. Such surfaces, as Adams discovered — and presented in his book with Olive Whicher, *Plant Between Sun and Earth* — are fundamental to the forms of living organisms. Thus water and its relationship to these special mathematically describable surfaces was to be investigated with regard to possible influences on the water. Water and surface are inseparable. Water is always moving over surfaces and, over shorter or longer periods, depending on the erodible nature of the material, the shape of those surfaces is influenced. Water in movement also always tends to create a multitude of surfaces within itself but these can be influenced by the condition of the water. In a living context water movement plays a major creative role in formative processes. Indeed every physical forming process is at some time in a fluid state.

Work in connection with the mathematical surfaces came to an end for the time being when available information was exhausted after the death of George Adams in 1963. My activities with the restoration of Rudolf Steiner's original models continued in Dornach, Switzerland, until 1970, but already in 1965 I received an invitation from Francis Edmunds, through an introduction by Rex Raab, to join Emerson College in England and develop the sculpture activities there. At the same time I began giving courses at the Rudolf Steiner Seminariet in Järna, Sweden, founded by Arne Klingborg. Thus until 1970 my time was divided between Dornach and Emerson College with annual visits to Sweden. During this period the courses I was giving demanded a renewed and intensified study of morphology and metamorphosis.

At the invitation of Theodor Schwenk I returned to the Institute for Flow Sciences in Herrischried at Easter 1970 for what might have proved to be some periods of longer duration. Without knowing what tasks were to be taken up I prepared myself with certain questions on which I saw the possibility to work and discussed these with Schwenk on arrival. Schwenk was very enthusiastic about the results of my investigations and encouraged me to proceed further. The main discovery I made concerns the possibility to generate rhythms in streaming water by means of a specific level of resistance. One of the first things I could demonstrate with this technique was the practicability of incorporating mathematical surfaces within a vessel system for the purposes of investigation. It had not been previously attainable — due to the dominance of gravitational and rotational forces — to encourage water to caress proffered surfaces intimately, by spreading out over them, or indeed following specific curves on them. The way was now open to continue with 'path-curve' research.

Preparations were made for further work periods during 1970-71 and I hoped for a continuation of this collaboration but lack of space and possibly the fact that I did not have a scientific training seemed to make it unsuitable to continue within the Institute. My wife and I finally moved to England to take up residence at Emerson College in Sussex, where while building up courses I could continue with my design research.

Acknowledgments

First and foremost I wish to express my indebtedness over the last thirty years and more to my wife Alfhild for her unfailing patience and support, together with Johanna and Thomas; then, in my professional life, to Rex Raab and Arne Klingborg; and in connection with research, to George Adams, Theodor Schwenk and Nick Thomas.

It has been Schwenk's work — demonstrated in his book *Sensitive Chaos* — to which I owe the inception of my own design and scientific research. His seminal contribution to the reawakening of a true modern consciousness for water is fundamental to the contents of the present book, which I consider in all humility as the description of one consequence of his investigative research.

Regarding the work itself I wish to mention many who have made a distinctive contribution in one way or another. Since the discovery and development of the Flowform method in 1970, its implementation has found expression with well over one thousand projects in over thirty countries through the activities of the Flow Design Research Group at Emerson College and about thirty associated individuals or groups in some twenty-five countries.

Arne Klingborg was one of the first to recognize the Flowform's potential and initiated the first major project in Järna, Sweden. For the ensuing project in Akalla, near Stockholm, Iain Corrin and Nigel Wells joined me. Nigel became the main contributor to design developments for the next ten years and has remained a most faithful friend and supporter out of his own endeavours in Sweden since 1985. From very early on Felicia Cronin helped us in all the different aspects of recording the work photographically. During this pioneering period especially, Martina and Christopher Mann proffered major support for protective measures such as patents and also design developments, and this has continued in one way or another up to the present.

For research and development a number of organizations have contributed over the years: Rudolf Steiner Wissenschaftliche Fond, Nuremberg; Cultura Stiftung, Heidenheim; Mercury Arts Foundation and the Margaret Wilkinson Fund, London; Helixor Stiftung and Fischermühle, Balingen. To all of these we are very grateful.

Nigel Wells carried out the Sevenfold I Flowform Cascade, followed by Hansjörg Palm with whom the Sevenfold II Flowform Cascade was developed during 1986, completed by Nick Weidmann who has remained my main design collaborator until today. Nick Thomas began his contributions during the 1970s with manifold scientific investigations and mathematical design advice in continuance of George Adams' theme concerning path-curve applications to water research.

Francis Edmunds, John Davy and Michael Spence of Emerson College Council supported the continuance of the work at Emerson from 1970 onwards, our activities always remaining completely self-financing. Our first attempt to build a Rhythm Research Institute in 1980 was supported financially by many people. Kersti Biuw for instance not only participated in our work but brought a major donation. Failure however to attract adequate funding at that time brought about a halt in the building. The next years proved to be the most difficult, during which time forces intervened that attempted to take the work out of our hands and claim authorship.

During the 1990s contributions for a building began to appear, increasing those assets accumulated through our own projects. Support from Rudolf Steiner Foundation USA, Unni Coward in Norway, Katrin Fichtmüller in Switzerland, Vidaraasen Landsby in Norway, Cadbury's Foundation, U.K., and finally Software Stiftung in Germany, all encouraged a reactivation of building plans and a move ahead. After various attempts, initial plans were created with Lars Danielsson in Sweden through which we were able to gain planning consent. We also had a number of offers of expertise and help to build from friends in far-flung places, to whom I wish to express my gratitude. As events have unfolded, Emerson College architects Nick Pople and later Tom Rowling carried the process through to the final stages within an overall programme of College expansion.

A number of associates, most of whom participated with the Flow Design Research Group for shorter or longer periods, have carried the work in different countries. Herbert Dreiseitl began with Flowforms in Germany and gradually built up extensive activities concerning water in town planning (see his *Neue Wege für das Regenwasser* and *Waterscapes*). Iain Trousdell carried out many Flowform projects in New Zealand, making a major contribution

From left to right: John Wilkes, Arne Klingborg, Walter Liebendorfer, Abbe Assmussen, Rex Raab.

Photo: Benjamin Boardman.

with Peter Proctor in the area of biodynamic farming. Mark Baxter as architect has upheld activity in Australia, gradually joined by a number of others in that country. Andrew Joiner came back from Africa to help with a number of important projects then went off to start his own operation in Yorkshire. Aonghus Gordon of Ruskin Mill took the initiative to start a Flowform production workshop which continues to supply cast stone Flowforms. In the USA, Chris Hecht and Sven Schuenemann continue with great vigour. Jenny Green began with great enthusiasm, but her work with Flowforms has now been passed on to others.

In Norway at Vidaraasen, Lars Henrick Nessheim built up a casting workshop and operation, supporting in addition a number of Flowform developments. Jürgen Uhlvund has taken over the workshop and is investigating the effects of water quality and timing on the casting process. Jörn Copijn supported many initiatives leading to extensive activity in Holland. Hanna Keis appeared in Denmark, to carry out numerous projects supporting the work for many years. Michael Monzies continues with work in France. Pit Müller some years ago opened his 'Wasserwerkstatt' in Dortmund, has carried out many technically efficient installations and generated a number of design research projects for us.

Thomas Wilkes has taken up the important development with ceramic Flowforms. Nick Thomas, Georg Sonder and Jan Capjon are contributing to mathematical solutions connected with my Virbela screw, on which Don Ratcliff also worked many years ago.

Others, all of whom cannot be mentioned, are or have been active in Italy, Portugal, Estonia, Iceland, Finland, India, Brazil, Switzerland, Germany, Greece, Spain, Canada, Israel, Belgium, South Africa, Kenya, Taiwan; and all the time new initiatives are beginning. Altogether we have experienced collaboration in over thirty countries, more recently also in Mexico, China, Poland and Hungary, with possible projects in Rumania, Turkey and Peru.

Due to this exposure of Flowforms in many countries there have appeared references, illustrations and articles in well over three-hundred publications, these are however only those I have personally heard about.

I am grateful to Robert Kaller in Germany for initiatives with colleagues such as Pit Müller to support the work and particularly at a certain stage in preparation of this book. My thanks, as well, to Herbert Koepf for his contribution to Chapter 1 on time and rhythm; to Mark Riegner for his collaboration on aspects of metamorphosis in Chapter 3 and Appendix 1; and to Nick Thomas for providing the appendix material on scientific and technical perspectives. As the writing comes to a culmination, I am greatly indebted to Costantino Giorgetti who has given invaluable help in finalizing the manuscript. He has also played a major role in planning future activities in connection with our Rhythm Research Institute and guaranteeing that the building has been able to go ahead. We thank Floris Books warmly for their dedicated contribution to the presentation of this publication and the German publishers Urachhaus/Freis Geistesleben for their patience in standing by the endeavour for many years.

In acknowledging the contribution of so many people involved, I naturally regret any oversights and hope that any inadvertent omissions can be corrected in future editions.

A. John Wilkes
Emerson College, Sussex, U.K.

Introduction

Water is essential to virtually all processes we can imagine whether natural or technological. Certainly we complain when there is either too little or too much — both can bring death. Balance is achieved just when there is not too little and not too much in any given context.

In many places water is still used primarily in moderate measure just to keep organisms alive. In our modern technological world however water is used in excess and too often only as transporting agent or energy producer. All such uses deteriorate its capacity to support life. We have taken advantage of the sea, of the rivers, the lakes and underground water reserves to satisfy our ever-increasing demands. Rachel Carson was probably the first to investigate thoroughly the consequences of our unsustainable activities. Facts about the consequences are becoming increasingly well known and it is not the purpose of the present book to dwell on this theme.

This book is directed to those people with an open mind who are interested in our environment and are willing to admit that it is in need of our active support and this, in its own terms. Such an attitude can simply no longer be considered a romantic notion. The situation is serious and humanity still needs the earth and its manifold species for some time to come in order to carry out tasks it has not yet fulfilled.

As long as we continue to consider nature, organisms and life as something merely physical, technological and chemical we are missing a comprehension of the whole picture. There are obviously much more subtle aspects which nature is trying to show us, if only we are willing to see.

This book is not making all manner of claims for a particular approach. It attempts to open up another attitude to water and indeed through this a more sustainable approach to all resources.

The real function of water is not merely to make things wet, absorb heat and excess waste, generate energy and provide transportation, all of which it can do in moderation, but something far more subtle. It is there to mediate to every living thing, the movements and rhythms of the total environment. This means the total embedding of each organism within the most subtle aspects of its surroundings. There is nothing living that can survive without this mediating wonder of water which sustains all relationships.

Water is the element of movement *per se*. Its function in nature is that of universal mediator. Everything living is inevitably dependent upon water which is itself the physical carrier of rhythm.

An hypothesis could be formulated thus: As rhythms are fundamental to life, would it be possible to apply an advanced understanding of them in such a way that water's capacity to support life be improved? Moving water is inseparable from surface — either inner or outer. It either influences surfaces over which it flows or is influenced by them, and it creates surfaces within its own volume. So rhythm and surface are components which have to be investigated together.

Is it thinkable, then, that rhythms in conjunction with specific surfaces could have an influence upon nature in a potentizing sense, thus helping to support healing and harmonizing processes?

Questions such as these led to the discovery of the Flowform — a vessel that can be designed in a multitude of sizes and shapes and is, by *virtue of its proportions*, capable of inducing rhythms in water streaming through it. Having discovered the Flowform — a name that was coined some years later during its development and implementation — it became evident to me that the rhythms induced may not only be of aesthetic and visual interest but also provide support for life processes.

How do rhythmical processes actually manifest in the physical world of organic forms? They appear as metamorphic relationships in the living organisms surrounding us. Metamorphic relationships have to do with physically discontinuous processes in which components nonetheless have a connecting link, such as the main plant stem which unites individual leaves. Processes such as these often demonstrate changing forms within an organic totality, in contrast to the changing form within a growth process that is physically continuous.

Metamorphosis is then to be understood as a dynamic formative process 'external' to the physical manifestation. As an example, *we can think of something vital happening* between one leaf and the next, which leads to a change in its shape. This is an essential aspect of the narrative relating to Flowforms as they are ultimately intended to provide water with an 'organ of metamorphosis' by means of which its life-supporting capacities are enhanced.

I would like to call what follows the 'biography of an idea.' It is not intended in any way to give an exhaustive treatise on water, a task carried out impressively by other publications, such as those by Theodor Schwenk, Callum Coats, Alan Hall and Charles Ryrie.

First of all we look at rhythmic phenomena in the natural world around us, and create a mood in which we can recall for ourselves what have been our experiences in this realm.

We then examine the origins and conception of the Flowform Method. The Flowform, though arising out of the artistic domain, has functional and scientific aspects as well as aesthetic associations, and these we attempt to explain, together with the scientific work on path-curves which actually preceded the discovery, development and application of the Flowform Method itself.

We go on to describe the wide range of applications, technical, social and aesthetic, which have allowed further exploration and discovery over the years. The first commissioned work was related to a water purification system into which rhythmical movements were introduced to support plant and animal processes. There followed a more socially oriented project within a children's recreation area on a high-rise estate. Other commissions followed, needing a variety of designs for features in parks and private gardens, often requiring individual solutions for specific circumstances.

The final chapters describe some of the consequences of these discoveries, recapitulate the achievements to date, and indicate our future goals.

PART 1

Rhythm and Polarity

1

Water and Rhythm

One tranquil early morning, I sat on the stony shore of a Scottish loch, far from the sea. The loch surface was as still as a mirror and nothing seemed to move, until I noticed the water slowly rising over those rounded stone shapes. The surface tension of the water's edge crept like mercury up the dry stones. Reaching the top of a stone the water would suddenly jump together, generating a band of gliding rings which expanded and faded away on the mirror surface.

All at once I became aware of the great in-breathing motion of the whole ocean surface, rising. This was a rare and unusual experience of the tide rising visibly, inexorably to the slow rhythm of the beckoning Moon. The land was drawing up its mantle of water. An in-breathing would be followed by an out-breathing and so this immediate experience led to a picture of its opposite, the releasing of the mantle. Later would come another process of gentle almost imperceptible withdrawal, without commotion, leaving the wet rocks emerging from the falling tide.

A gentle breeze crossing that placid surface reminded me of the quicker rhythms of ripples and waves. A single rock protruding above the surface intercepted the gentlest of waves, and from this interference proceeded the most wonderful spiralling wave patterns, swinging out to the left and right.

Then, as I moved along the shore, the sound of a movement came from a trickling stream entering the pool. At first the stream held its own, entering the still pool of water in a straight line, but very soon through the surrounding watery resistance a meandering rhythmical swing appeared, only to disappear again into chaos and mixing as momentum was lost.

From just this one occasion, it was clear that the subtlety of movements that can be observed in water is virtually indescribable. Already, I had observed three very different patterns of rhythm in water, patterns that we shall come to examine more closely. Following

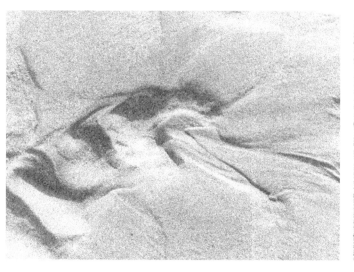

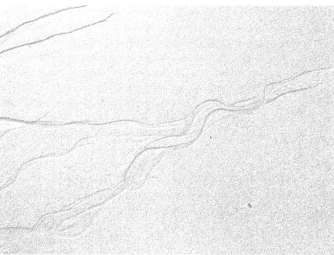

Figure 1.1 (left). Sand forms of a particular kind created by water flowing from right to left, draining an area down a slope, then eroding the sand and in a rhythmical form.

Figure 1.2 (right). Sand forms revealing the 'history' of the resulting meander, firstly a straighter channel due to greater flow, gradually swinging out more and more as the flow decreases, the crossing points can be clearly seen.

Peter Schneider (1973) we may call them laminar, harmonic and turbulent. Where for instance a small spring rises under water at the edge of a lake or on the coast, the flow-rate, tending to vary, will show up differently in the moving grains of sand. First the standing wave of a circle can accommodate the flow. Increasing, the water suddenly flows out of one side of the circle changing the radial symmetry to a bilateral symmetry. At this point the system begins to pulsate. A further increase in flow brings turbulence. As the flow changes, so the phenomenon moves back and forth through the three states. It is fascinating to see how rhythm appears between laminar and turbulent just as the delicate state of bilateral symmetry is established.

On almost any sandy shore, we find that water rhythms leave their patterns or forms in sand left by a retreating tide, or where a stream crosses a beach. On tidal coasts a multitude of forms can be found, from the hard ripples of many beaches to the most sensitive relief impressions of all manner of plants and creatures, a collection infinitely richer than the most fanciful of imaginations could dream (Fig. 1.1).

From the smallest stream of water a signature remains in the sand, in which the whole history of a meander development can be read (Fig. 1.2). Where the flow is strongest, the path is relatively straight, then as the flow decreases and the sand resists its passage, it swings in wider and wider alternating meanders to left and right, but through static crossing points until the flow stops altogether.

As these ever-present phenomena display, any experience of

water will usually lead us to one of rhythm and it is surely the medium through which rhythm is most readily expressed. Yet of all the commonplace descriptions about water as it moves and forms the face of our planet, how many relate to its rhythms?

This is where we begin.

Rhythms, which are fundamental to the creation of all manifest forms in our surroundings, are carried by fluid processes for which water is basic. As water moves over the earth, however, it appears to lose for a while its rhythmical order due to the multiplicity of forms over which it flows.

Rhythm appears by virtue of an interplay between gravity and, let us say, levity; or contraction and expansion, pressure and suction, centre and periphery. The evidence of these alternating movements appears everywhere in the outer world, from planets moving across the fixed stars to the grains of sand formed by wind and water on the beach. Nature's rhythm embodies stronger and weaker elements of repetition and regularity but seems never to repeat itself mechanically. Always an aspect of change emerges. Every movement and every form is essentially unique.

Rhythms are manifest, too, in the living cycles of the natural world. The very fabric of our cosmos is rhythmical and its influence finds a way down to the minutest events in nature. Life processes embody this quality of rhythm so innately that we experience them as inseparable. Life is manifest rhythm, and as such is a microcosm mirroring the macrocosm.

Creatures of land, sea and air live very intimately with their environment and it is only human beings who have increasingly, though not entirely, emancipated themselves from nature's rhythms. Not only this, but gradually the knowledge of such connections has diminished, while the intervention of technology has served mostly to intensify a separation from natural rhythm. So much so that these cosmic relationships are largely ignored in modern culture, except where experiences like jetlag teach us humans that our body functions and, in the long run, our very wellbeing, are not entirely independent of Nature. Ultimately, all creatures live within the breathing of the earth-sun rhythm of day and year.

Living creatures lead their lives in cycles, rhythmical time structures which underly growth, reproduction and activity. Recent research into these phenomena has been pursued within a fast expanding branch of the life sciences which we call chronobiology.

Many of the rhythms in a living organism originate in its inner order, and are known as endorhythms. The organism may also respond to cues in the immediate environment, perhaps biological or meteorological. Not all influences are local, however. Some periodicities are directly in phase with the sun and the moon, inviting us to form a colourful image of earthly-cosmic relationships of life.

To mention one example of these cosmic forces: farmers and biologists know well the importance of day-length for the development of crops. So-called longday plants require a period of days with more than fourteen hours of light if they are to round off their stages of growth. Shortday plants need fewer daylight hours for the same maturing. Here it is the duration of light itself which is important, not light intensity or other factors such as temperature. Light is the source of energy for photosynthesis, but over and above this it acts as a formative force in the time-body of many organisms. From the daily and seasonal rhythms of light, a broad spectrum of effects emerges, including the seasonal development of plants, and the activities, reproductive cycles and migrations of animals.

Equally in marine forms of life, we find many diurnal, monthly or seasonal rhythms, influenced by tides, temperature, brightness during the night or other external moon-related factors. However a wider, qualitative aspect of lunar influence suggests itself, the modality of lunar rhythms appearing through the relation of the moon to the watery element as it flows outside or inside plants and animal organisms.

We have come to believe, through study and observation, that an understanding of rhythms will enable a true science of nature to develop. Whether alive or inorganic, the spatial objects which we perceive in the manifest world are all impermanent. Forms, positions, and other characteristics, all change or vanish in time. There is continuous movement in nature, however fast or slow. Our understanding of what we see grows more real and complete if we understand how it has come about and where it is likely to go according to what we know of its past. The time-scale involved extends beyond, in both directions, the gusty movements of the wind, the eddies in the river and the age of the drift block laid down during the Ice Age.

In its own way every species is specifically interwoven within its earthly and cosmic environment and history. These surroundings vary rhythmically in time according to the movements of earth, sun, moon and stars. There is an ideal harmony and regularity, too, in the movement of the planetary bodies. While the irregularity of meteorological conditions interferes with this, nonetheless the life of earthly creatures continuously responds to this interplay between terrestrial and cosmic realms.

Through recent work on Gaia theory, the concept of the Earth itself as a living organism is once more gaining credence, as it surely had in older times. The idea can now however be understood in more comprehensive and precise ways. In such a framework, we may think of the water cycle as equivalent to the planet's blood circulation. Water is the 'blood of the Earth,' a concept coined by others earlier in the last century (see Schauberger and Steiner), and as such we may see it as rhythm and flow.

Without water there would be no movement in nature, and without movement there would be no life. The water cycle, from source to ocean to atmosphere, constitutes great rhythms within which multitudes of organisms exist. This vast circulation of water in all its states and movements — solid, fluid and gaseous — exists largely to fulfil the task of mediating rhythms of every kind within the total environment. The inner world of every living organism carries the memory of this watery environment from the gateway of birth. As we shall see, from that moment on, the organism is sustained by rhythms which, if they fail, take the very gift of life itself away with them.

2

☙

Rhythm and flow:
the water cycle

Leonardo da Vinci made the remark: 'Movement is the source and cause of all life.' This conjures up particularly an image of water. First, then, let us evoke a picture of the water cycle and its movements as it progresses from source to ocean. The term 'precipitation' describes well the linear descent of raindrops, almost totally under the influence of gravity. These may land among hills or mountains where they continue to move. Every drop wherever it falls tends towards others creating tiny streamlets. At this start of their journey, the slopes may channel the streams into vigorous, narrow, fast flowing torrents that cut deep trenches and chasms, and tumble in the thrust of waterfalls. Then as the slopes become less steep and flatten into valley landscapes, the streams gain in size, becoming wider, slower flowing and proportionately shallower rivers which meander in generous curves. Gravity comes to have less and less dominion and the watercourse comes eventually to relative rest in lake or ocean.

When we contemplate such movement in the created world we may ask how it comes about. Movement, we can say, only arises through the workings of polarity. Polarity in its many aspects is the basis of physical existence, and without such oppositions the behaviour of material things as we observe them would not be possible. Contraction is unthinkable without expansion. If only one of these aspects existed, all would be either extraordinarily compact and immovable or completely invisible and non-material.

It is in the nature of water to take on whatever shape is offered, identifying itself with the earth beneath while its free upper surface relates to the total globe. The behaviour of water is thus always expressed between polarities. Water moves over sloping ground due to the influence of gravity, but this can only happen because it has already been lifted up by means of a completely opposite process. This movement over the Earth's surface and eventual evaporation away from the Earth's surface are aspects of the water cycle, itself the rhythmical expression of a fluid circulation which maintains life on the planet. Meandering rivers are the heart of the whole cycle (Fig.2.1).

When considering the total cycle we realize that the river holds the central position in a sevenfold process in which we clearly see polarities at work: contraction, expansion; gravity, levity; straightness and circularity. Initially water vaporizes from all surfaces and assumes for a while a state of vast expansion during which it is totally invisible to us. Under the right conditions it again becomes visible in cloud formations in various layers of the atmosphere. This is the beginning of a process of attraction towards the earth during which gravity takes over. Multitudes of growing spherical droplets descend in such a benign way that water is able to reach down again to the very surfaces from which it has gently risen. The great confluence of tributaries begins from every spot where a drop falls, this gathering of waters forming the third stage of contraction within the whole cycle. Now, following the straight-line aspect of precipitation and descent, the flow reaches lower ground and we find the innate tendency to form alternating spirals, out of which the meander is created. On approaching the ocean the river often broadens out in the delta area and breaks into a number of arms, due to slowing down and deposition of load. Thus the expansion process is enhanced, followed by discharge into the ocean from which evaporation again takes place on a large scale.

As already mentioned, a number of investigators have spoken of water as the 'blood of the earth.' This image becomes increasingly apt the more we consider the earth as an organism. What is the pulsing meandering river other than an organ which is indicator of the health of the land?

Consider first the heart itself as a regulative sense organ which registers the state of our health. If our blood vessels were to become blocked, straightened, cut, pressurized, diseased, or loaded with poison beyond a certain limit, we would not survive. Circulation would cease, thus bringing the heart to a standstill.

Already in the 1880s William Morris Davis saw the rivers and

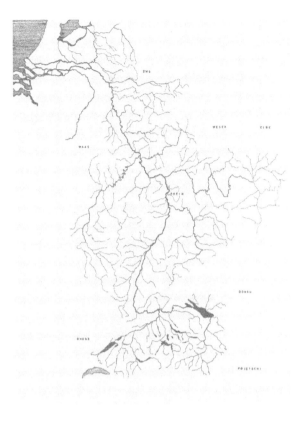

Figure 2.1. The drainage basin of a great river system, the Rhine.

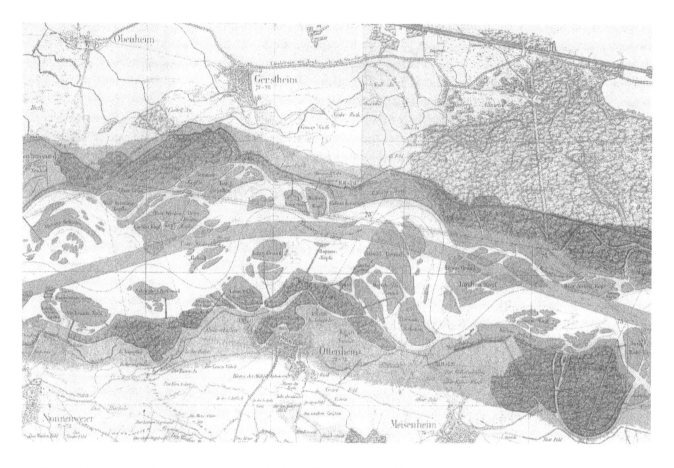

Figure 2.2. A section of the Rhine as it passes through the Black Forest. The map was drawn in 1836 when long deliberations were taking place regarding the control and straightening of the river.

their valleys, too, as living organisms that grew from infancy through youth and maturity to old age. Yet the reality today is that many rivers are in chronic disorder and are in need of urgent remedial treatment, precisely because they have been blocked, straightened, cut, pressurized, diseased and loaded with poison beyond the limit of their endurance.

The first symptoms of disorder appeared at least a century ago with actions taken to straighten major rivers in order to make them more navigable and useful for commerce. This 1836 map of the Rhine passing through the Black Forest, shows the original path of the water flow, with the intended navigation channel superimposed (Fig 2.2).

But in undertaking these huge engineering 'improvements,' no account was taken of how the river's rhythmical life processes would be attacked. No audit was made of how, through this straightening process, the river would lose its existing functions in relation to the flood plain, the regulation and maintenance of

flow-rate and groundwater levels — all of these managed naturally and efficiently by means of the rhythmical meandering of the water flow.

Figure 2.3. Erosion caused by the Alsek-Tatshenshinin meandering river in the Tongass National Forest, Alaska.

Photo: Getty Images/Stone.

The meander

The meander is the 'pulse' of the river (Fig.2.3). If we observe the streaming of water and closely watch the movements caused by flowing under the influence of gravity, we see that through the slightest resistance, curving vortical forms are generated. Water always tends towards the sphere and, in streaming, towards the vortex. Reciprocating left- and right-handed movements inherent in all streaming processes lead to meandering erosions whether in minute sand forms, rocky slopes (Fig. 2.4) or mighty rivers. We find the very same patterns in ancient stone formations in which fluid movements have become fossilized (Fig. 2.5).

The meander of the river, then, is due to the actual liquid nature of the processes themselves. Naturally diverse conditions, including geological influences, interfere with what would otherwise be regularly developing forms, as we see in the phenomenon of the path-of-vortices (see Fig. 2.6).

The meander, like the entire water cycle, must be seen as a phenomenon which occurs as the balancing out of opposite tendencies. The straight line and the circle are forms out of which respectively the meander is born and into which it finally dies. From straight-line gravity-bound movement water finally rests in the horizontally static circle, as for instance in the so-called oxbow lake which separates off as the main stream joins up to bypass it.

The following table (Table 1) gives an overview of some of the rivers' main characteristics throughout its length.

Figure 2.4 (above). Such meander channels can be found below glaciers where streams have caused erosion over thousands of years.

Figure 2.5 (above right). Fascinating corrugations within rocks of different composition reminding us of their fluid stages.

From asymmetrical to symmetrical forms

There is another opposition that strikes us whenever we observe the behaviour of water. It has to do with the difference between the effect of a stream and that of a thrust (see Chapter 4). Wherever a streaming process is involved, asymmetry appears, owing to the fundamental law that whenever a liquid encounters resistance it will curl into a curve.

Water is:
invigorated   and eventually   *degenerated*

Water moves through channels:

steep	becoming	*horizontal*
narrow	becoming	*wide*
fast	becoming	*slow*
straight	becoming	*curved*
deep	becoming	*shallow*

Consequences

Water is:
moved by gravity   *settled by gravity*

It changes:

from oligotrophic	to	*eutrophic*
from cool	to	*warm*
from oxygenated	to	*deoxygenated*
from clean	to	*polluted*

It has a tendency towards:
erosion   *deposition*
fixed channels   *shifting channels*

It shows:
lack of nutrients   *overloading with nutrients*

Table 1: The meander is born of a straight line — and dies into the circle.

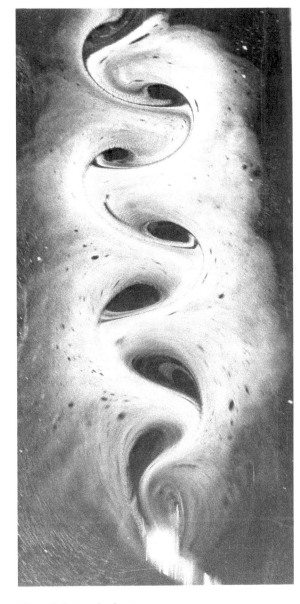

Figure 2.6. A path of vortices.

Figure 2.7. A broken dyke on the coast reveals the radiating pattern of sand deposition.

Figure 2.8. A wave pulse pressing through a passage in a sea wall shows the formation of a symmetrical double vortex. Photo: Norbert Didden

However a thrust or specific impulse can lead to symmetry as we show with the following illustrations. A broken dyke releases a load of sand carried in the water to form an overall symmetrical pattern (Fig. 2.7). Here a wave thrust through a gap or a hole creates a typical mushroom-like head (Fig 2.8, see also Fig 4.14 on pp. 55f). This example shows how the thrust is resisted from all sides evenly by the surrounding water. In Fig 2.9, can we imagine how this relatively symmetrical erosion has come about in the Colorado river?

Nearly every manifestation of movement in water is asymmetrical. Only when such a form is mirrored will symmetry be established. It is interesting to compare living forms in this respect. An organism will manifest a thrust as it is physically conceived, like an egg or a jellyfish. When a thrust persists the resulting stream creates meandering forms like tentacles streaming out behind. A mushroom bursts out of the ground and presents us with the well-known symmetrical form to which it lends its name. The

vertical thrust tends to generate a circular, radial symmetry form in the vertical axis. When observed from the side, bilateral symmetry normally appears.

A most interesting example which I look upon as a kind of archetypal plant form is the seaweed *Sacchariza Polyscheides* (Fig. 2.10). In its natural environment it is born out of an intimate relationship to water movements. It has a hollow spherical (head-like) organ out of which radiate linear root forms in all directions, followed immediately by dynamic rhythmical meandering surfaces which gradually fade away along opposite edges of a flat stem. This in turn splits up into numerous flat branches. So first we have a spherical symmetry out of which lines radiate, then the rhythmical middle region where bilateral symmetry appears, followed by a relatively symmetrical branching which nevertheless increasingly succumbs to the chaotic movements of the surrounding water. Once more the threefold pattern of laminar, harmonic and

Figure 2.9. An extraordinary formation in Colorado; one imagines the water must have been oscillating back and forth laterally over a very long period at full spate. Photo: Carolyn Bean Pub.Co.

Figure 2.10. This wonderful seaweed called Sacchariza Polyscheides *is for me a real archetypal kind of plant, at home of course in the water. At both ends complex asymmetry, one expanding into flat fronds, the other radiating linear roots from a spherical-like organ. In between a transitional realm along a flat central stem with edge corrugations developing into a dramatic meandering form either side, bringing a tendency to bilateral symmetry in this rhythmical sector. The plant often wraps its roots round a pebble which leaves it free to move with the sea, but weighs it down so that it can actually travel in a vertical posture.*

turbulent, revealing the intimate relationship between water behaviour and organic forms.

Some years ago I formed the idea of developing a fuller comparison of the river or catchment area and the plant, as represented by the tree. Many thoughts arose through this exercise which could certainly be improved and extended but I include it to encourage the reader to develop further ideas (see Table 2 on p. 36).

The sacrificial character of water

In a technical context, in contrast to the natural world, water can be manipulated purely as a utility. It can be turned into a deadly agent of destruction by virtue of its weight and movement, the heat it can accumulate, the poisons it can absorb even down to electrical, magnetic and radioactive content. On the other hand this same substance can be employed as the gentlest ministrant to the most tender blossom.

Even those potentialities, for good or ill, can be ignored where our interest is narrow and utilitarian. Where we merely require its weight and momentum, water can be made to move more or less as a solid block with very little or no disturbance in its volume, as for instance when it falls through a duct to drive a turbine.

Water can also, through human design or in its natural state, rest in a tank or cavity, almost if not completely without movement. It can be to all intents and purposes devoid of intercourse with the surroundings, unwritten and amorphous in its inner structure. In a natural setting, this condition can bring an experience of something pristine, heavenly or spiritual, 'out of this world,' perhaps in a still dark grotto where water is hidden and secret as though waiting in suspense.

But the moment water is released from these imposed conditions its volume can be penetrated with membranous movements. Both on its surface and within its volume, movements in response to influences of all kinds reveal a rhythmical potential. This whole world of movement in a body of water is normally closed to our direct perception and experimental devices have to be used to make visible this secret realm of nature.

Theodor Schwenk is one researcher who made great efforts to reveal these mysteries and in the book *Sensitive Chaos* some of his findings are shared. The very title of the book, taken from a comment of Novalis, underlines an aspect of water which is of vital importance: its extraordinary sensitivity. This character can also be implied in the word *sacrifice*. Offering itself totally to its surroundings, water sacrifices itself almost to the point of losing its identity. Just as Goethe speaks of the 'deeds and sufferings of light,' so can we speak of the deeds and sufferings of water, the 'mediator' *par excellence*. Water cannot resist the actions put upon it because it is essentially open for this purpose and function. It therefore cannot reject any adulteration of its integrity.

Only we human beings, through an enlightened consciousness, can protect water by preventing unsustainable use, unnecessary pollution, careful management of technology and development of a better understanding of its precious nature. It is towards these ends that we continue our exploration into the inherent character of water. Having studied a few examples of the formative principles of rhythm in outer nature, we now move to the principle of metamorphosis which is an aspect of rhythm as it manifests itself mainly in the living world.

The river	The tree
Rivers are dammed or drained.	Trees are lopped or felled.
Ground water is regulated by the meandering river.	Trees help to regulate the climate, the water content of the atmosphere.
Fish and many other species live in and around the river system.	Birds, animals and insects live in and around the tree.
Breeding upper reaches	Seed production periphery
Water courses encourage vegetation to create shadow and reduce evaporation.	Trees grow leaves that encourage transpiration and evaporation.
Turbulent water in upper reaches (oligotrophic) takes in oxygen.	Foliage gives off oxygen.
Rotting detritus in water gives off methane and other gases (eutrophic).	Living foliage absorbed carbon dioxide gases.
Asymmetrical tendencies.	Symmetrical tendencies.
Weight of water falling is used for energy production. A concentration process. The sun has caused the water to evaporate then fall, over a short cycle.	Timber is used to generate energy by burning. An expansion process. The sun has helped to build the substance of the tree, over a long cycle.
Form of the river bed is made through water's erosion of the surroundings (destructive).	Water supports the building of the tree in a creative process (creative).
Bacteria and algae participate in photosynthesis and form a water-born food chain.	Light converted to energy by photosynthesis in the plant cells as basis for the food chain.
In past ages glaciers were formed at the periphery of the river system, where it actually begins.	In past ages coal was formed as a fossilised product of the tree growth process, as an end stage.
White, light.	Black, dark.

Table 2: A comparative view of the river and the tree.

The river	The tree
The river flows from the periphery to the delta.	The tree grows from the trunk to the periphery.
Its confluence of tributaries gradually builds the main stream.	The trunk divides up into branches and twigs growing outwards.
Dominantly two-dimensional.	Dominantly three dimensional.
Dominantly horizontal.	Dominantly vertical.
Flows in every direction of the compass over the face of earth (two-dimensional).	Points in every direction from the centre of the earth over its whole surface (three-dimensional).
Water gathers in increasing volume.	Fluids disperse into the periphery.
Picture of gravity	Picture of levity
Flow stability	Form stability
As water increases in volume in the main more horizontal bed the meander tends to become more complex. Eventually it separates into arms in the delta.	The individual roots come together at ground level to form a straight vertical trunk. As the volume of material decreases the relative path of the branches and twigs becomes more complex.
Volume of water flows in the river bed down to the sea where it evaporates.	Sap rises from the roots up the trunk to the periphery into which it disperses.
Carries material from the periphery to the sea where it is deposited (This is also abused in all manner of ways).	Sap carries nutrients etc. from the ground via the stem out to the periphery where water is released in a pure state (transpiration).
From the ocean mineral forces move upstream under the river bed, to the furthest tributaries, bringing minerals to the soil for the plants.	From cosmic spaces substance is taken in and flows peripherally downwards under the bark in the cambium, helping to build the tree.
Harvests are gathered and give a measure of the quality of health in the system. Organisms are plucked out of the water (middle reaches).	Harvests are gathered which give a measure of the quality of health in the organism. Fruits are plucked off the tree (upper reaches).

3

Metamorphosis

Dynamic processes

Metamorphosis means 'change of form,' and as a general definition we may say that it concerns the relationship between two or more physically separate forms in sequence within an organic whole. It has to do with diverse and changing relationships, between or among elements of any given totality or organism, associated usually by dynamic processes which are not of a physical nature.

Each organism reveals to us that it has a life of its own, as well as its specific relationship to the larger natural context of which it is a member. A familiar example would be that of a plant, with its root, cotyledons, leaves, blossom, stamen, seed and fruit, the relationships of which already exhibit different types or qualities of metamorphosis.[1]

A plant develops in time, hence it will never present itself fully to our physical eyes at any one moment. It is either a germinating seed, or it forms stem and leaves, or its cycle declines following the seed or fruiting stage. The shape of a preceding leaf along the stem, is not the cause of the form of the leaf that follows, in the sense of plain causality. Both are members of a rhythmically appearing, consistent sequence of transforming shapes. They manifest the wholeness of the organism. They display similar, yet at the same time discretely different, shapes. Only in thinking can we accompany the plasticity or fluid nature of those transformations of forms; in principle they are all one and the same. The growing plant produces a rhythmically appearing sequence of the finished results of continued transformation.

Today our comprehension of phenomena is all too bound by an emphasis on the material, physical aspects of existence. In the following discussion, therefore, rather than remain within a narrow definition of metamorphosis, I intend to take a wider, freer view of metamorphic principles as a way towards exploring artistic and scientific forms which embody a relationship between physical and spiritual dynamics.

Observing phenomena

Taking the first steps towards understanding a language of movement requires a reorientation in observing natural phenomena. It was Goethe who pioneered a dynamic approach to nature observation on which the present work is based. In this approach, an isolated phenomenon holds little value in itself; only when it is interrelated with other forms does its significance emerge. The aim is to search for an invariant principle amidst a family of diverse phenomena, in other words to apprehend *the unifying movement among changing forms*.

We begin then by considering aspects of time and change. Our awareness of time arises from appearances. In daily experience, objects appear in sequence, say the carriages of a long freight train which are passing by. We might say, as we observe them, that it took quite a bit of time until the last carriage disappeared. Thus through movement we become conscious of time. Yet time is not a thing, but a measure. We measure in minutes, hours, days, by following the apparent path of the sun from dawn to dusk, and the movements in space of the clock's hands. We relate time periods to our subjective experience as we describe them as fast or slow.

As we discussed in the preceding chapters, when like events occur in like intervals — say, a sound is repeated, drops of water fall from the gutter, the surf of the sea swells and falls back — we speak about repetition, or perhaps rhythm.

Where repetitions of like appearances occur in all kingdoms of nature, including the inorganic, they often leave the evidence of continuity. The pattern of small ridges in the sand along the beach; the herring-bone figure of some cirrus cloud; the accrued layers of silt in what was once a diluvial pond, are more or less transient or lasting spatial forms of inorganic matter.

We observe the result; of the preceding process we perceive segments only. The continuity of the linear or rhythmical process we form by thinking. Through the activity of thinking, the fractionated details of one percept are moulded into the unifying notion of the process, the essence of which thus becomes conscious. In then imagining or anticipating further activity in the process, we shift mentally from 'what has become' to that which is 'becoming.'

A proper concept of time sheds light on the relation between being or essence on the one hand and appearances on the other. Facts may be related to each other by dint of their very natures. If according to this relationship they follow each other in a succession, that is, the earlier one must be there and give way to the later one, then time arises. Awareness of the inner relation between events in time becomes the threshold from the level of their appearances as events to their underlying being, or idea.

This becomes particularly evident in the kingdoms of living nature, where the connection between being and manifestation, in other words idea and appearance, is very different from that of the inorganic realm. With movements in living objects human thinking is concerned with a different task as the manifestations of the successive events often have the appearance of discontinuity.

This is where Goethe introduced the principle of the 'archetypal plant.' By meticulously observing and interrelating botanical forms — whether the successive forms of different plants or the various organs of an individual plant — a unifying principle emerged. This ideal element he called the archetypal plant, the common 'idea' which stands behind the changing forms, the general within the particular.

This archetypal plant is neither a static blueprint nor a fixed notion, but a cognitive activity to which our habitual mental categories do little justice. By weaving different botanical forms into one metamorphic sequence, it is possible to experience, through thoughtful contemplation, a dynamic unified gesture behind the phenomena.

This ideal principle can be grasped as a living reality which unites all plants regardless of their diverse morphology, and can be considered the essence of 'plantness.' We recognize both an oak tree and a dandelion as plants, but rarely pause to consider just how we are able to do so. Our recognition is characteristically associative. If we attempt to penetrate through our wall of conditioned associations, however, and remain patiently with the observed phenomena, we then approach a region in ourselves where we can meet phenomena in a new, enlivened way.

Pursuing the analogy of the metamorphosis of the plant in relation to water, my search turned towards the 'dynamic unified gesture' that might be discovered in the phenomenon of flow.

Metamorphosis and water

How, then, to create 'an organ of metamorphosis for water'? This was the key question which precipitated my research into the rhythms of water over thirty years ago. The development of ideas on definable types of metamorphosis, has been my main interest through an extended study of morphology (see Appendix 1). There is a need to create a threshold and key which gives entry to this very complex realm of experience. An intimate knowledge of metamorphosis will be increasingly important for artistic and scientific evolution as we deepen our understanding of development processes, especially where there is physical discontinuity. While growth processes are physically continuous, metamorphic processes are physically *dis*continuous and this encompasses the issue we have before us.

As we have seen in discussing the water cycle, polarity is fundamental to all physical life on earth (see Chapter 2). Without polarity no physical manifestation of a created world would be possible. If we take just expansion and contraction, for instance, all physical phenomena have a relation to both, in some proportion. The presence of both is essential for life. If only contraction were active, all existence would be infinitely compact and invariable so that any life forms would have been impossible.

It is therefore the continuous adjustment of proportions between opposites such as concave and convex, expansion and contraction, heat and cold, wet and dry, that generates movement. Movement, ever-present and continuously reiterated, thus provides the source for all forms whether living or mineral.

Potentially within this repetition of movement, rhythms appear in the fluid processes of external Nature, only to break down again and again when not held within a living entity. Life processes, however, exist within a consistent rhythmical framework in the absence of which they cannot remain manifest. Rhythm is the maintenance of a particular pattern of movement, albeit one which is ever shifting, but usually within given limits. Organisms embody such patterns building up their physical existence, in time. The organism maintains its identity but follows a changing rhythmical pattern, thus developing physically a metamorphic order. Thus polarity is also the originating universal factor which relates all things from a metamorphic point of view.

In his book *Sensitive Chaos*, Theodor Schwenk describes the result of a straight-line movement through still water, revealing what are called 'paths of vortices.' He terms the phenomenon a *Wirbelstrasse*, and illustrates it profusely in his work. Paths of vortices occur wherever streams of water pass each other or objects stand in the way. In nature's watercourses they are a familiar sight when vigorous enough to influence the modelling of the water surface and even gentler movements can be detected where the sun's rays cast their shadows on the stream bed.

In experimental situations we can both create and observe vortex sequences in such a way as to reveal the typical characteristics of metamorphosis already described above. Later we shall illustrate that such water patterns display an ordering polarity with a distinctive resemblance to plant growth. As this phenomenon constitutes the main inspiration which led to the discovery of the Flowform method, I should like to begin the next section by describing in more detail the approaches that have enabled me to observe and work experimentally with paths of vortices in water, as well as other behavioural patterns.

PART 2

Discovering the Flowform

4

∞

Experimenting with water

From drop to waterfall

Every day we could do little experiments with water, but we rarely take the time even if we have the opportunity! How often have we watched a tap dripping or running, thinking 'I must fix it,' at the same time noticing the tiny agitated trickle whipping back and forth across the glazed hand-basin down to the hole. The first thing we observe is that wherever there is a slope, water moves down it, never to return in the same composition. The steeper the slope, the narrower and faster the flow, while the more horizontal the gradient, the wider and slower the flow.

In the beginning was movement, and all forms are born of movement, from the humblest waterdrop to the majesty of a waterfall. At every level, there are realms rich in changing shapes that can be instructive to observe, and for any reader who so wishes, the following simple procedures are easy to carry out.

Playing with streams

To go a step further in our water play, we can allow a stream to flow at varying rates over a non-absorbent surface like a piece of glass or perspex. Using a length of plastic tube or flexible drinking straws we create a syphon to deliver a stream of water over the glass or perspex from a vessel placed at a slightly higher level. Raising and lowering the vessel can modify the flow-rate very subtly. We discover a play between the effects of gravity and resistance. Changing the relationship between gradient and rate of flow produces results between a straight line and curve. The greater the slope, the stronger the flow-rate, the straighter the stream. Reduce either and the stream whips in a meander.

The pulsing puddle

Now, a notable stage is provided by the condition where the form of the meander becomes stable and static, in other words it no longer whips back and forth. The form of the meander created by the moving water is held within its own surface tension and within this 'skin' water streams imperceptibly through. In order to make the movement of water visible a few crystals of permanganate are placed at strategic spots.

In many different ways through such streaming systems, we see oscillatory or rhythmical processes appearing. In playing with the static form, by stretching a bend for instance, into a bulbous puddle contained still by the water's own surface tension, it is possible to generate a relatively stable pulse created by the resistance of the liquid within the little puddle reservoir to the passing stream. It all has to do with proportions. So the pulse is achieved by creating just the right amount of resistance to the flow. Such a system of proportions is in fact regulating the flow by introducing a rhythm. This generally will not readily happen in an open natural system, if only for the reason that too many influences come into play. It is only where a higher organizing function is involved, as in the experiment with the experimenter or in a living context, that the vessel generates a pulse. This phenomenon can have some bearing on the idea that the heart's activity is not causing the blood flow but first of all resisting and thus regulating it (see further Chapter 6).

To summarize: the phenomenon demonstrates the creation of a form held by the liquid, which then in turn regulates the flow. As we shall see later, organs are created by fluid processes which are themselves then regulated by the organ.

In both the above experiments it may at some point be necessary to treat the surface with wax or a suitable polish to prevent the water from developing a more intimate relationship with the surface, when it will tend to spread out, as though the surface is porous. We find that the moment water is released over an absorbent surface, the whipping meander is metamorphosed into a complex of wave rhythms, each selvedged with its own corrugation.

The water drop

Each drop of water is unique in the moment of its existence and will never have that identity again. The contents of that momentary drop will disperse into all its neighbours in an instant, in a series of falling drops a rhythm often being observable, the drops joining together to form a pulsating stream.

A drop of coloured water falling a few centimetres into a glass of stirred water will disappear as soon as it touches the surface. But if the enfolding water is quite still and movement is only present in the falling drop as it penetrates the surface, a quite orderly process of mixing can be observed taking place. However, the exact character of this process depends upon a number of factors, and this is where experiment comes into its own. The process will be influenced by the size of the drop and the height from which it falls; its temperature compared with that of the water in the glass; the weight of the colour in the drop which makes the otherwise invisible process visible (a slightly warmer drop can counteract this weight); and any turbulence within the drop itself. Although probably too quick to follow with the naked eye, the familiar 'crown' of drops we have seen elsewhere photographically illustrated, forms in response to the moment of impact, and rhythmical waves emanate over the surface as this subsides. The drop can however enter the volume of water and maintain its identity now as a vortex ring, and this vortex ring will spawn a number of further rings, creating a circular pyramid of arcaded forms.

Veils or sheaths

The remarkable thing revealed here before our eyes, is that water is moving in veils or sheaths and these are maintained only while a certain degree of movement exists. As the movement is resisted on all sides more or less evenly by the still water and is brought to a standstill, the form gradually disperses and disintegrates into an amorphous mixture. The original coloured drop loses its identity as its momentum is eroded and nullified.

To reveal this process on a larger more exciting scale, the following simple experiment can be done. Take a large syringe casing, lower the open end into the water and inject a little colour through the tiny hole now at the top. Putting a finger over this hole, raise the cylinder so that the coloured water inside it is above the outside water surface. Release the hole, so that the water falls in the cylinder, and the plug of coloured water descends as a vortex ring more or less intact, dependent upon its momentum.[2]

Let us try something else. Stirring several litres slowly in a cylindrical glass vessel we can inject colour from a syringe into the centre. Immediately out of the cloud of injected colour, sheaths assemble in the vortex, rotating with the whole volume. Continuing to watch we see an unexpected vertical rhythmical displacement of the coloured water, strongest in the centre, diminishing to the periphery with the decreasing rotation. As the overall movement slows down, so do the forms disappear.

Erosion forms

In a large flat sand box many different arrangements can be made to observe how moving water creates form within an erodible material such as sand. The moving grains provide a very sensitive medium to respond to all kinds of formative processes. There are many parameters with which one can experiment, from the grain size — single or graded — to gradient and different streaming processes. In order to preserve any particularly exciting result, it is possible to spray the sand carefully with a quick-setting lacquer, in order then to make a plaster cast by applying a mix very gently indeed. Naturally, without fixing the forms in such a way, the slightest influence of another medium will destroy the original effect.

The waterfall

Finally a rather more hidden phenomenon concerning movements that develop within a waterfall under particular circumstances. Even though it may be difficult to see, one can try some experiments and attempt to observe what happens. Many different flow-rates from varying heights can be observed. It will probably be necessary to take fast exposure photographs to catch what occurs. Just as the extremely delicate and sensitive stream within a volume demonstrates three regions of laminar, harmonic and turbulent, so can a waterfall. The photograph taken by scientists at the research body SINTEF in Trondheim University, shows hidden forms in water falling from 3-4 metres. As it falls, the air forms a resistance and a momentary rhythmic process ensues before breaking down in a chaos of droplets (Fig. 4.1).

It is of course possible to experiment on a smaller scale. Small metal channels can be made, perhaps 10 cm across, out of the end of which water can be allowed to fall. The edge can be cut in different profiles, from an edge which is at right angles to the flow direction, to one which is diagonal to the flow. Or curves can be cut, over which the water falls. This thin sheet of falling water, and the spatial forms it takes up, can be observed. How long can this sheet of water be kept intact? One experiment consists of following the curves at the water's edge and bending wire to 'sit' in these edges. These wires can for instance then be carefully drawn apart horizontally, to observe how far the sheet of water can be 'stretched.'

The path of vortices

As we have seen, living forms constantly strive for a balance within the harmonic realm. Wherever streams of water pass each other or objects stand in the way, we find the phenomenon of vortex sequences (*Wirbelstrassen*), a common sight in natural watercourses (see Chapter 2).

The controlled creation of vortex sequences has been used for a long time to study and record these behavioural effects experimentally, and investigative work in this area led me to the explorations resulting in the development of the Flowform.

In an early experiment, the vortex sequence can be observed in a shallow tank lined with black plastic and containing a mixture of water and glycerine.[3] An extremely fine powder is dusted on the surface of the water. We normally use lycopodium, the spore of the club-moss plant, which behaves somewhat like sand, in that the particles move relatively unhindered over each other. It is dusted on the surface of the liquid before it is moved and shows up against the black background of the vessel, revealing the surface whirls caused when an object is passed through the liquid in a straight-line movement.

There are a number of parameters which have to be carefully adjusted in order to obtain optimal results. To begin with, the application of the lycopodium — difficult to repeat exactly even when carried out mechanically — influences the result obtained. It can be dusted as evenly as possible over the water; it can be concentrated on one spot, along the line of motion or either side of this line.

Using only clear water, no movement is visible in it. Various indicators other than lycopodium, such as oil-based colours (used in the well known marbling techniques), reveal different qualities of definition. Metal powders can also be used, and as they do not float on the surface, they reveal other influences of the movement

Figure 4.1. A highspeed photograph by SINTEF Trondheim University, of a falling stream of water, about four metres high. We see this extraordinary phenomenon of air resistance creating the harmonic condition of rhythm within the waterfall which then collapses into turbulence below. Above one can observe the laminar straight-line condition.

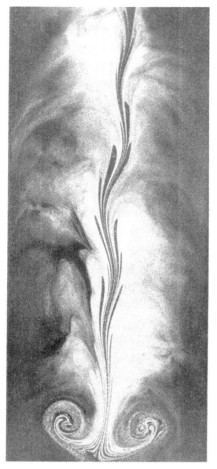

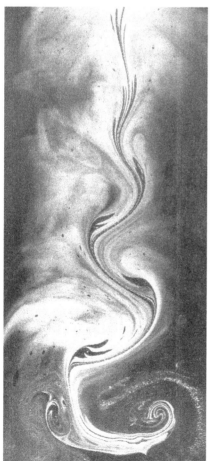

induced. By increasing the viscosity of the water with syrup or glycerine, the movement is slowed down and even encouraged to come to a standstill, and this allows adequate time to study the result. The depth of the liquid influences the degree of momentum carried within its volume by the straight-line movement which itself can be varied in speed and executed with objects of different width.

It is a matter of experimenting with all the parameters until the optimal results are achieved: the viscosity, the depth, the speed of movement, the width of the object used, the method of lycopodium application. With the liquid too thin or too deep, the resulting forms will not come to a standstill; too thick or too shallow and not enough movement is carried in the liquid.

The straight-line movement can be slow or fast: from a standstill where no form is evident, to a very slow movement where a slightly meandering form is achieved. The meander is 'born' out of the straight line. In contrast, a fast movement will lead to a complex riot of circling vortical forms. Somewhere in between will be found, with each set of conditions, the optimal speed which relates best to the other parameters of viscosity, temperature, depth, and width of object. This will produce the most harmonious rhythmical picture with a number of alternately moving left- and right-handed vortices. I like to call this a 'vortical meander.'

To look at specific results, we can take two particularly apt examples from Schwenk (see *Sensitive Chaos*, Plates 29 and 30.) (Figs. 4.2, 4.3) The first indicates the result of a slow movement, the second a slightly faster movement made under very similar conditions, which allows for a comparison of the two. The first demonstrates a gentle meander. One can see from the very form of the second that the movement has been more vigorous. What is fascinating here is that with a little imagination one can discern something like the picture of a plant. At the top of the picture or where the movement starts, there is naturally very little momentum and the potential vortices cannot develop further than a slight meander. As the movement proceeds, horizontally of course, but here shown vertically and moving from top to bottom, the left- and right-handed vortices are progressively able to develop further. In other words, they curl inwards more and more. The lycopodium indicator is stretched further into the vortex movement at each stage. We can see that each 'vortex' has its independent 'growth' but at the same time has its 'place' in the total movement complex or movement organism. The forms achieve a specific maturity before

FLOWFORMS: THE RHYTHMIC POWER OF WATER

they come to a standstill. This also depends upon the parameters described above. If for instance the liquid is too deep, movement is likely to continue too long and the picture will be lost.

If this example is indeed to be seen as representing an impression of a plant, the picture must display the typical characteristics of leaf metamorphosis. We are reminded here of the metamorphic tendency to move upwards, through stem-building to spreading, membering and lancet or sprout-like forms in the upper leaves, proceeding in reverse order to the leaf growth process itself which begins with a sprouting and develops through membering and spreading to stem building.[4]

Another remarkable thing is that this 'upright' plant picture is produced by a movement from the top where the movement starts, to the bottom where the movement has gained momentum and then stops. This initiating, formative process proceeds in the opposite direction to that of its 'growth' from below upwards. This gives us a little insight into the extraordinary secret that the formative metamorphic processes entering from the periphery inwards are steadfastly consumed and incorporated by the growth processes developing outwards with the growing plant itself. The more subtle peripheral energies of the Sun are absorbed by the physical proliferating substances of the earth.

Another picture follows of an actual example of Malvae leaves (Fig 4.4) so that the patterns can be compared.

The next example demonstrates the result of a stronger movement leading to a dramatic development of the vortices (Fig. 4.5). A wider object drawn through the liquid tends to create a progressively stronger development in the vortices. If a very narrow object is used, like a pencil, the vortices tend to remain parallel in development (Fig.

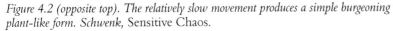

Figure 4.2 (opposite top). The relatively slow movement produces a simple burgeoning plant-like form. Schwenk, Sensitive Chaos.

Figure 4.3 (opposite bottom). The quicker movement begins to transform the meander towards what may be called a vortical meander. Here the vortices are beginning to look more like leaves, the lowest one with stem, the blade spreading and already membered. Progressing 'upwards' they become simpler, typical for the leaf development in a plant.

Figure 4.4 (top right). Although this leaf process within the Malvae looks different, in principal the typical form activities are expressed from below upwards, stem building, spreading, membering and sprouting (Bockemühl).

Figure 4.5 (bottom right). Still more energetic movement allows the path of vortices to develop further into a more complex metamorphic sequence.

Figure 4.6. Using a very narrow object pulled through the water, creates a train of parallel vortices. This picture demonstrates also that the slightest imperfection in the movement brings irregularity to the result. Water is extremely sensitive to any minute influence.

Figure 4.7. An example showing now a great similarity to the formation of an animal embryo.

 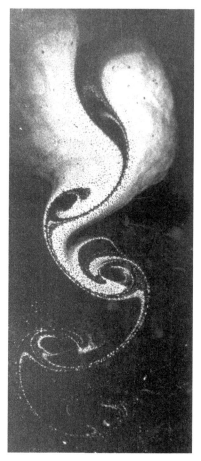

4.6). Here it can be noted that the slightest irregularity in the straight-line movement or indeed in the surrounding water has its consequences in the pattern of the vortices. We see that diagonal to the applied straight-line movement alternately from left and right, liquid is sucked inwards successively, often, more or less at right angles. Due to the diagonal movement these mushroom-like forms are somewhat asymmetrical. A single such detail (Fig. 4.7) looks something like an animal embryo. Above is a snout-like form with large eye, below the forelimb and further below the hindquarters with tail.

The next two pictures show, on the left the process still in movement, taken also with a relatively slow exposure (Fig. 4.8). Although the phenomenon is completely flat the result gives the impression of a strongly three dimensional form. The moving object can still be seen at the lower end of the picture. A few seconds later another photo is taken, showing that the three dimensional impression has

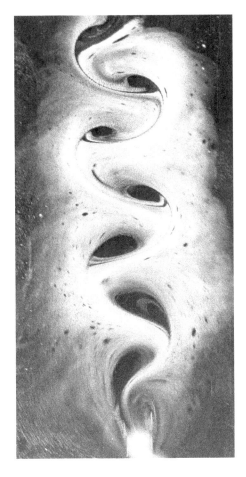
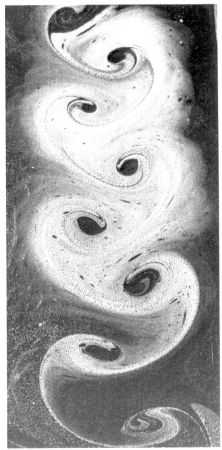

completely disappeared and a more graphic result is seen (Fig. 4.9). At this moment everything has come to a standstill. Once again we discern how intimately form is connected with movement.

By dusting the lycopodium in two parallel lines, the movement can be carried out between them. The indicator is thus drawn inwards creating a kind of negative picture which can be particularly dramatic due to its simplicity (Fig. 4.10).

With the above conditions maintained it is possible to create a double effect tending towards the symmetrical, by using two straight-line movements through the liquid, adjacent and parallel but with a shift, that is the second object a little later than the first. Again it is necessary by trial and error to find the best relationships to achieve an optimal result (see Fig. 4.18, see p. 57).

Even in this relatively simple experiment one experiences the extraordinary sensitivity of water. It is extremely difficult to achieve

Figure 4.8. Taking a photograph at 1/30th sec. while everything is still moving — the object can be seen below — creates a strongly three dimensional impression where the phenomenon is absolutely flat on the surface of the water.

Figure 4.9. This three dimensional impression disappears the moment everything comes to a standstill. A photograph taken a few seconds after the previous one.

Figure 4.10. An example created by dusting the lycopodium in two parallel lines either side of the movement. The white is drawn inwards from the periphery only.

a perfectly even picture exhibiting the ultimate potential which lies in the phenomenon. The slightest irregularity introduces some kind of malformation in the result.

Three-dimensional experiments

In the two-dimensional type of experiments described above, we detect principally the harmonic realm in the world of movement. When we construct experimental conditions involving three-dimensional flow, the laminar realm becomes observable before the harmonic.

If we observe the flow of coloured water into a tank, with just the right flowrate laminar flow can be demonstrated in which no change of form or mixing takes place, as the stream is held together by its own momentum. For this experiment, we use an aquarium-type tank or a long container of at least ten or more centimetres depth. We fill the tank with water and allow it to settle, at best for an hour or two. Surroundings should be at as even a temperature as possible. A plastic tube is arranged as a syphon with the lower end fixed horizontally in a cramp below the water surface, and feeding from a smaller reservoir at a higher level containing coloured water. This feeder tank can be raised or lowered to control the flowrate very accurately. Practice makes perfect. It is always a matter of not-too-much or not-too-little momentum.

The temperature relationship between the incoming coloured stream and surrounding water is critical. In order for instance to compensate for the weight of the colour, the incoming stream must be very slightly warmer than the surrounding volume. The aim is to achieve a stream which moves horizontally between the extremes of rising to the surface or sinking to the bottom. It is in this sensitive realm of movement that environmental influences, even planetary constellations can show a change in the form.

Soon an oscillation will occur due to the surrounding resistance, which slows the stream, tending towards a momentary instability. Rhythmical forms appear, clearly showing harmonic interplay between flow and resistance. This is a significant and sensitive interim condition between laminar and turbulent, which may not always appear and when it does, may even be ignored by the observer. If the flow is too fast, turbulence will set in right away, leading directly to formlessness, and no sensitive interplay takes place.

When the full sequence is allowed to evolve at an appropriate tempo, the rhythmic condition is followed by a gradual cessation of

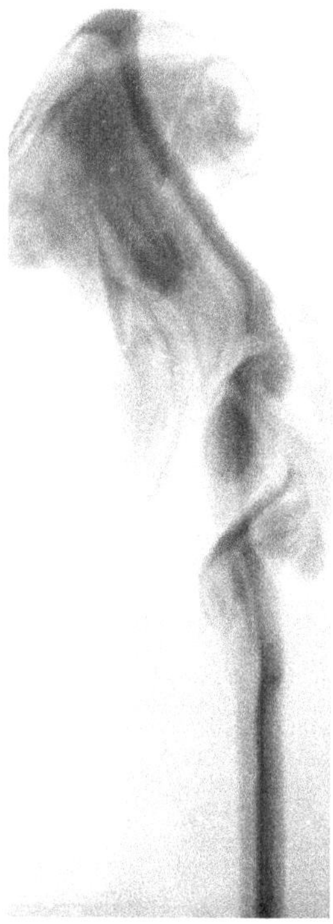

Clockwise:

Figure 4.11. A stream of coloured water entering a tank of still water which has preferably been left to settle for some hours. The stream needs to be very slightly warmer to counteract the weight of the colour. The stream has to move horizontally.

Figure 4.12. Another example showing a very delicate process from a laminar condition, through the harmonic to the turbulent. Form is held in balanced movement — not too fast and not too slow. As an expression of balance between opposite tendencies the harmonic realm is that in which the whole of nature's forms appear.

Figure 4.13. A result where we seem to see limb bones with a joint in between.

54

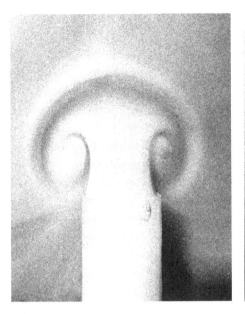
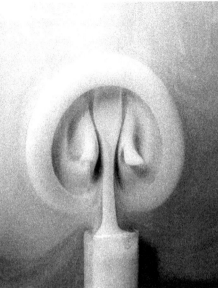
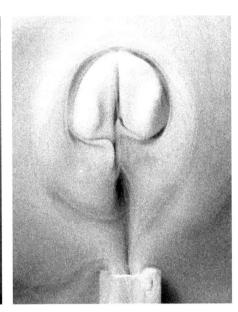

movement with a disintegration of form through gentle turbulence to amorphousness.

The three states — laminar, harmonic and turbulent — are revealed in Figure 4.11. First the laminar, that is, the straight-line realm of no-form out of which the meander is actually born; then the harmonic where ordered formative processes can take place.[5] Finally the turbulent, the realm of excess form where order no longer governs or is hidden in complexity. Figure 4.12 shows somewhat more detail within the system. In Figure 4.13, a somewhat more irregular form gives an impression of long-bones with a joint in between.

An interesting variant on this experiment — and a very exciting discovery which arose during a seminar demonstration — can be carried out as follows. If the incoming stream is colder than the water in the tank, it sinks to the bottom and forms a layer of colour there. Moving a narrow object through this, in a straight line across the bottom of the tank, creates a path of vortices as previously described. However, due to the volume of water above, each vortex can develop also vertically.

In all these instances, there can be no one ideal specimen but only a tendency in the direction of an archetypal metamorphosis. There is a more or less even transition between opposites, between the straight line — out of which the vortical meander is 'born' — and the 'circling' vortex into which it 'dies.'

Figures 4.14, 4.15 and 4.16. Symmetrical forms produced as the result of a thrust rather than a stream. Introduced through an orifice into a shallow tank of water, a gentle thrust of water mixed with a little clay slip or colour is allowed to move horizontally. In this situation typically symmetrical forms will be produced in contrast to streaming water where asymmetrical forms appear.

Figure 4.14 shows the beginning of such a process.
Figure 4.15 reveals a further development a few seconds later.
Figure 4.16 shows how the slightest influence in the very sensitive condition causes asymmetry in the gradual process of slowing and disintegration.

Photos: Felicia Cronin. Experiment: Philip Kilner.

Asymmetry and symmetry

Figure 4.17 (a- g). With this realm where asymmetry dominates I questioned how one might be able to observe a symmetrical complex of movements. This is a photographic experiment. Left and right negatives of the same picture are gradually superimposed. In the central position a quite vertebral-like picture is produced as a metamorphic sequence.

Figure 4.18 (opposite top). Another experiment, using two objects drawn through water at the same time, parallel but one displaced further back so that the two meanders would combine to form a symmetry.

Figure 4.19 (opposite bottom). The previous natural result where the two sides become easily dissimilar due to the extreme sensitivity of the experiment, is shown here exactly symmetrical by mirroring one side.

All these examples of the 'path of vortices' are, due to their very nature, asymmetrical. Indeed water movements in general lead to asymmetrical forms, and it is somewhat rare to find symmetrical examples in horizontal movement. However, symmetry does occur in vertical movement. For instance, the waterdrop itself when suspended in the air, tends towards the spherically symmetrical. When falling it oscillates horizontally repeatedly through the radial symmetry of the circle while becoming bilaterally symmetrical (for instance ovoid) in the vertical axis.

In the horizontal, only a single thrust produces a symmetrical form (Figs. 4.14 – 4.16, and compare Fig. 2.8 on p. 32). When sustained as a stream it becomes again asymmetrical. The dynamic of the streaming asymmetrical forms gives way to the somewhat more set single-thrust symmetrical forms found universally in nature.

All living forms tend in some degree towards forms of symmetry, whether spherical, radial or bilateral. It is as though the 'asymmetrical' freedom inherent in fluid processes has to be restricted 'symmetrically,' in the realm of the living. Becoming aware of this dichotomy of asymmetry/symmetry, awoke in me an intense interest which led to further investigation. Suffice a couple of examples.

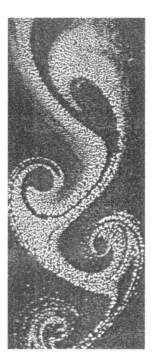

A row of pictures (Fig. 4.17) illustrates the gradual superimposition of a path-of-vortices, mirrored photographically, thus producing symmetrical forms. The central position is not unlike a row of vertebral bones, the adjacent examples reminiscent of metamorphosing organic forms.

Figure 4.18 shows a complex path-of-vortices produced with two straight-line movements, one somewhat behind the other, creating a 'shift.' On the top the 'natural' outcome, with almost symmetrical result; below, one side mirrored to illustrate a totally 'mechanical' symmetry (Fig 4.19).

These illustrations were a result of my exploration to discover a transition from asymmetrical to symmetrical forms. We speak of the left- and right-handed; forms mirrored lead to symmetry. Although this predisposition appears everywhere in nature, it is probably safe to say that the asymmetrical 'freedom' is never totally banished.

Preoccupation with this same theme prompted me to ask the question: what would happen if symmetry were 'offered' to water? The outcome of this question is taken further in Chapter 5.

5

Discovery of the Flowform Method

Working towards the Flowform

For a number of years during the sixties my main occupation involved making architectural models and carrying out sculptural restoration. For periods I was also developing and presenting sculpture courses in England and Sweden. This necessitated an intensified study into the morphology of botanical, anatomical and fluid phenomena, some of which has been described earlier.

Interdisciplinary work of this kind led me to ask fundamental questions regarding movement. All forms appear to be created through a relationship to movement and fluid processes. At some stage every rigid substance is in a state of flux. In this condition movement also leads to rhythm and rhythm to metamorphosis.

I had no idea of where I was being led in my searches but was simply open to developments, and the question that occupied me constantly was: 'Would it be possible to find out more about the genesis of form?'

In this area, I had already been fortunate to work with George Adams, a mathematician whose seminal research concerned Projective Geometry and the theory of space and counterspace. It was early in the 1950s that I first met him in connection with a thesis I was preparing at the Royal College of Art. He spoke about modern Projective Geometry, also termed descriptive or synthetic geometry, and offered to give me instruction. His work also examined in depth the mathematical phenomenon of the path-curve surface in natural forms. It is important to look briefly at this phenomenon here, as path-curve research continues to inform our work (for a fuller account, see Appendix 3 and Further Reading).

First described in mathematical terms by Felix Klein and Sophus Lee in the nineteenth century, path-curves were considered by

George Adams to be closely related to organic forms. He believed that such forms were associated with forces operating in nature which were non-gravitational and more related to suction and levity, the principle of expansion and growth. With Olive Whicher, he explored the evidence of such activities and influences in the plant world. Subsequent work by Lawrence Edwards over many years, studying forms such as eggs, buds, pine cones and organs such as the heart and the uterus, has clearly established the relationship of these ideal forms, when present, to the health of the organism in question. The closer the organism builds to this ideal form, the more vigorous and healthy it is, while, from an external point of view, unhealthy forces influencing the plant or organism will produce a deformation from the path-curve ideal. Adams' vision, as I understood it at the time, was that if we could as a first step allow water to experience such an ideal surface, would that water then acquire a higher potential to build organically, to support organic life-processes? It was Adams' idea to allow water to caress such surfaces, to spread out over them so that it could take up into itself something of the quality of this movement in space which is intimately connected with living things. We must remember that nature is doing this constantly; as inquisitive artists and scientists we must ask the reason why and for what purpose.

In early 1961, I began the task of constructing a number of path-curve surface models conceived by Adams. He wished to see these models built so that a basis would exist for future research into the applications of his theory in practical life. I became involved in this work in London, but it was only towards the end of 1962 that I was able to join George Adams as his assistant at the newly founded Flow Research Institute at Herrischried in the Black Forest. My task consisted of developing methods for creating apparatus based on path-curve surfaces designed by Adams, for the purpose of allowing water to pass over them. [6]

After March 1963, due to Adams' untimely death, the main thrust of the intended research into his ideas had to be postponed. Although the mathematical models of the surfaces based on his research could be completed with the information available, it had not yet been possible to obtain satisfactory experimental results on their effects. Difficulties had shown up in attempts to relate the flow of water closely enough to the surfaces in question or in moving it, even along preferred curves on those surfaces. Either gravitational or rotational processes would dominate. It also became evident through Drop Picture tests (a sensitive method devised for registering subtle qualitative changes in water) that the influence of materials used could override any positive effect achieved by movement.

I remained intermittently involved in this work for some years after George Adams' death. Then in early 1970, Theodor Schwenk invited me back to Herrischried to continue in some way with our investigations. I was not sure how that decision came about or really what he intended, but felt that here perhaps was an opportunity to follow up the earlier lines of inquiry. On March 24, I arrived in Herrischried again.

Due to my intervening work elsewhere over the years, I had formulated a number of thoughts and questions that could lend themselves to practical investigation, and I arrived with these in preparation for my initial conversations with Schwenk. They were recorded in my notebook as follows:

> A vocabulary of movement should be developed. How does water move over objects of different shapes — convex or concave — and how does it fall freely over an edge? Can it be introduced to a metamorphic process within a symmetrical context? — could this be sevenfold as in the vertebral column, which demonstrates a procession between expansion and contraction from head to tail?[7] Normally water moves quite freely and without order, symmetry or metamorphosis due to the fact that it has to respond continually to a multitude of influences. When it is allowed to respond within the context of a single straight-line movement, protected from all other influences, the resulting path of vortices (*Wirbelstrasse*) can exhibit a fundamental metamorphic sequence. Symmetry is here not evident or indeed generally in water movements but only appears in quite special circumstances.
>
> *Can an 'organ' be created that would enable water to manifest its potential to metamorphosis, where forms are generated in space and time?*

These issues were discussed with Theodor Schwenk on my arrival, and he gave me free rein to follow up my own investigative programme based on questions surrounding metamorphosis, symmetry and asymmetry in relation to water movement.

Background to the Flowform Method

As the experimental work continued, questions such as the above had been formulated and the observations carried out with open expectations and with no inkling of the outcome. On such an open-minded approach the whole character of the work was based — and continues to be based — and not on any relatively abstract hypothesis that had to be proved. The validity of the approach must lie in the character of the processes themselves and the way one works with them.

As already mentioned, water flowing over a complex stream or river bed exhibits a cacophony of rhythms appearing and disappearing at every instant. When I actually succeeded in obtaining such a rhythmical response in the controlled context of the experiment, it was generated by a set of suitably related proportions achieved without specific prior knowledge or intent.

This very result seems to validate the approach. Rhythm is the signature of life and appears wherever a natural living harmony of a special order is at work. It is as though this Flowform method plucks out of an infinite potential, the specific. It enhances the inherent tendency to vortical movement and rhythmical oscillations.

Although water moving in river and ocean is permeated with rhythms, due to multiple influences and water's sensitive response to them there is rarely, if ever, any clearly discernible order. But perhaps this is only apparent chaos.

Explorations

Water in its normal liquid state moves by virtue of gravity, out of the vertical along any degree of slope until it is absolutely horizontal, when it eventually comes to rest. As mentioned earlier (see Chapter 4), a falling sheet of water creates beautiful curves which can be varied extensively by changing the horizontal profile over which it flows, and this was one theme that I initially explored. Such waterfalls can be stretched between wires curved in space, an approach which could have been pursued much further, also

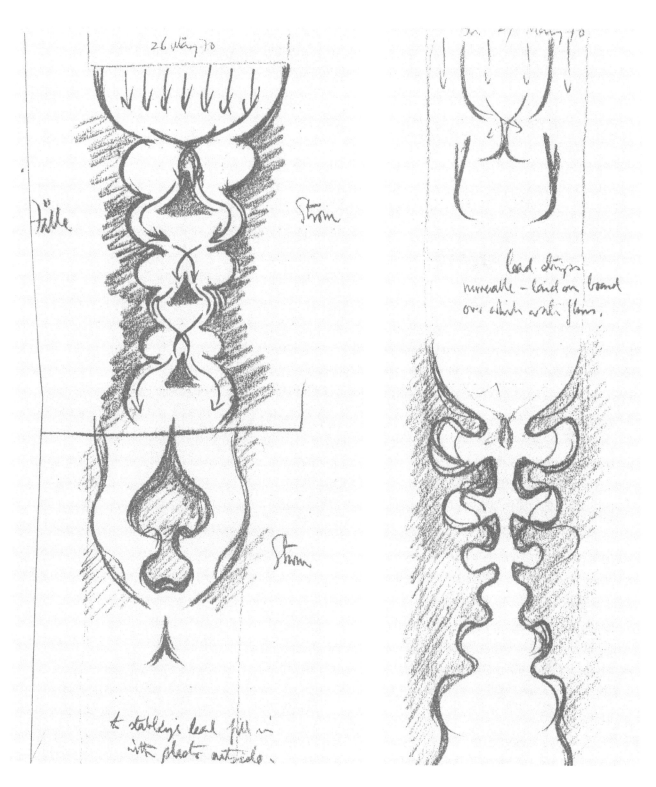

FLOWFORMS: THE RHYTHMIC POWER OF WATER

mathematically. However, after initial observations this approach was abandoned and other directions were taken.

I wanted also to study column-like forms with convex and concave surfaces in vertical section which would have been turned on a lathe. To what degree would water cling to such forms? Without getting completely into this line of inquiry, the planning led me more towards the channel idea. So, turning from the vertical expanding and contracting section, I considered forms that were increasingly horizontal. Indeed I transformed this column idea into a channel form.

The next step was to offer forms to water that would enable it to flow relatively freely but in an ordered situation, while not being disturbed by arbitrary influences such as sandbanks, boulders and the like. Perhaps a small regular freely falling stream of water from a running tap exhibits one of the rare instances where symmetry appears, again in rhythms of contraction and expansion. What would happen if a series of symmetrical forms were offered translating this theme of falling water to the more gently sloping context?

As already noted, my observations of nature had led me to the conclusion that water tends generally to move asymmetrically. When it enters the living organism, however, it helps to build symmetrical forms which in turn also depend on water for their manifestation. There is almost certainly no living organism that can exist without this watery process.

My line of questioning arose from the sense that the symmetry generated within living forms could arise from the condensing of a dynamic fluid process into an increasingly static and mineralized physical form.

The earliest drawing of a design introducing symmetry somewhat along these lines (26 March 70) shows the plan of a channel, expanding and contracting in rhythmical sequence (Fig. 5.1). When water falls from an edge it invariably draws together and meeting, spreads out again at 90° (see p. 56). This process as we know repeats itself a number of times. The channel that was to be made attempted to express this as a sloping symmetrical form with curved sections guiding the flow via crossing points on its flow path.

The second drawing (27 March) (Fig. 5.2) shows a meander mirrored about a central flow path. The narrow central channel thus opens out into cavities of changing size, presenting a systaltic

Figure 5.1 (opposite left). Ideas developing out of the phenomena observed. 26 Mar '70
Figure 5.2 (opposite right). The first plan of a channel idea.

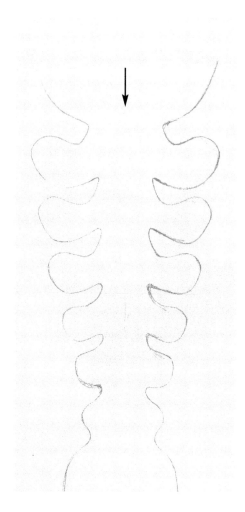

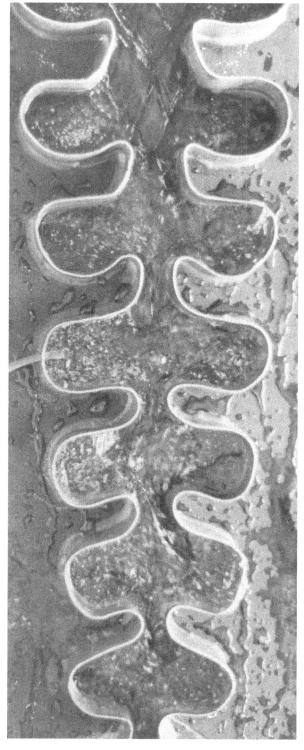

Figure 5.3 (above). A plan of the first actual channel made of lead strip walling on a glass base. Although the form does not betray anything special, water moving through reveals the hidden secret of proportion which leads to rhythmical oscillations, in the fourth cavity.

Figure 5.4 (right). A second channel with a tube inserted to measure rise and fall in the vessel.

Figure 5.5 (opposite top). Another type of channel to observe water movements.

Figure 5.6 (opposite middle). Experiment with cavities gradually becoming asymmetrical but nevertheless functioning. 27 cm x 70 cm.

Figure 5.7 (opposite bottom). A more dramatic asymmetry.

FLOWFORMS: THE RHYTHMIC POWER OF WATER

process — alternating between concave and convex. It was interesting to experience that the fundamental phenomenon of contraction and expansion was at once exhibited in the widening and narrowing of the channel.

On April 1, I built such a channel on a sheet of glass with ten centimetre high walls of sheet lead strips (the flexibility of which enabled changes to be made easily) sealed at the base with a special wax (Fig 5.3). The first cavity had an overall width of 26 cm, the fourth 17 cm. Water passing through on a gentle slope flooded the cavities and rotated, driven by the central stream of water. But the next moment provided a very unexpected event. Only in the fourth cavity, a vigorous lateral oscillation appeared.

Nothing in the design could have indicated such an unexpected movement, before the water itself revealed it. The oscillation had quite evidently to do with proportions of a specific order, achieved unintentionally, but nonetheless establishing the capacity to produce lateral pendulation in the flowing water.

The appearance of this pulsing phenomenon out of nowhere proved to be a key moment of recognition. The shape and proportions of a cavity give no clue as to how water might respond when flowing through it. This secret is revealed only by moving liquid.

The waterfall issuing from the channel rose and fell in a double rhythm, possibly a well-known phenomenon but to me at that moment a fresh and stimulating experience. A week later, I made another channel and at an appropriate place introduced a small tube, in which rising and falling water could be measured (8 April, Fig. 5.4).

This amazing phenomenon of a rhythmic pulse appearing in the stream of water prompted further questions. 'Could this effect be in any way similar,' was the question I was prompted to ask, 'to the condition within an organism where fluids such as blood are regulated rhythmically?' Is a rhythmical signature within an organism isolated by means of the specific proportions within that organism in order to maintain a desired influence?

It became quite clear with experiments on a number of variously shaped channels (Figs. 5.5–5.7) that it was the proportions which created the pulse. I coined the phrase, 'the not too much and the not too little' — for instance, apertures had to be not too wide and not too narrow. Symmetry had in fact no specific part

to play, but an instability was created through just the right degree of resistance and the right shape being offered by apertures, in combination with gradient and flowrate.

These conditions were by no means always easy to find, since there is no way of telling what should be altered in order to achieve a desired result. Once the resistance had caused the stream of water to be held up enough to bring about a deviation to left or right, a freer path into a lateral cavity presented itself, and the swinging process itself became stable. The continuance of this depended nevertheless upon fine tuning, the shape of the cavity also assuming quite some importance.[8]

In a number of further experiments carried out in small scale, I then began to see other extraordinary results. Now the oscillating stream flowed from a single channel alternately into a parallel complex of channels providing many cavities in which water could oscillate. In the larger final cavity of this complex the weight of water took it round the cavities giving for the first time the impression of a germinal lemniscatory or figure-of-eight movement (Fig. 5.8).

The increasing evidence for this pulsing lemniscatory movement represented a truly extraordinary moment of recognition which has significantly influenced the rest of my life. Here was a phenomenon which, once recognized, opened up almost, one could say, infinite possibilities for investigation and application.

During this exciting period from 1 April to 10 April, 1970, my experiments ranged between waterfalls and channels. All manner of combinations were tried out on both themes. Rhythms were measured, channels compared, holes introduced into the base, asymmetrical vessels investigated.

Experiments with asymmetrical channels opened up all manner of opportunities which would be taken up later, for instance as single cavity Flowforms. Exaggerated cavity shapes produced some interesting possibilities (Fig. 5.9).

I found that all parameters had to be fine-tuned to create an optimal movement in any one situation. Small vessels can generate a simple lateral wave rising and falling in the cavities but do not readily allow for a circling movement round the cavity. As above it was noted that rather larger forms with increased weight of water could lead to lemniscatory movement.

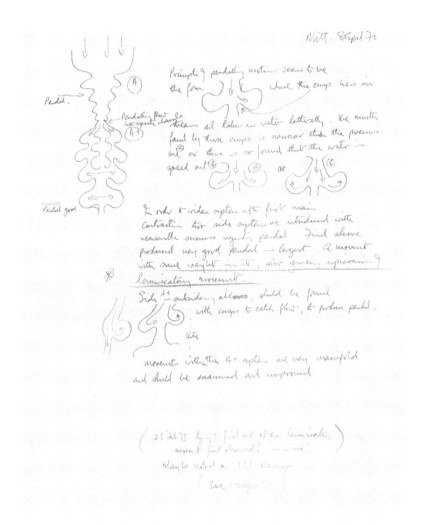

Figure 5.8 (left). Working with the idea of a complex of cavities combining symmetry and asymmetry. At the lower end of the channel (top left) a strong pulse develops, at the same time showing a germinal figure-of-eight movement.

It also became clear that the asymmetry could be taken so far that one cavity could disappear resulting in a single cavity Flowform which generated a simple pulse rather than a double one.

Figure 5.9 (above). Experiment carried out to test ideas, greatest width is 28 cm. Oscillations were possible in all cavities. Correct proportions were achieved by trial and error. This was still during the first week of investigations.

Resistance and rhythm

Resistance was seen to be the fundamental condition which leads to rhythm within the organism, perhaps a universal factor.

The next move was therefore to enlarge the whole vessel so that more freedom of movement was offered — and a flow path could develop rather than simply an oscillating wave to left and right — all relative to the normal viscosity of water.

Another factor became evident that inhibited the function once set in motion, namely that the oscillation itself changed the overall conditions and could lead to a cancellation of the movement through flooding. So this also had to be taken into consideration in

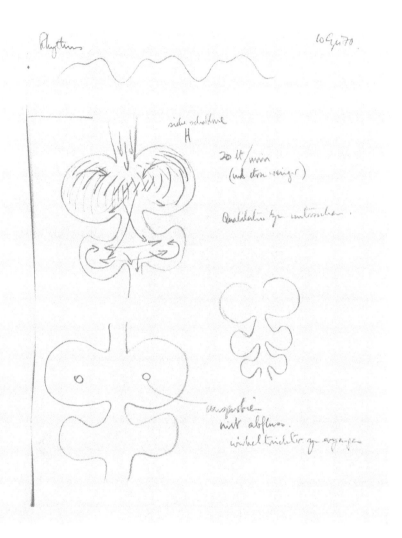

the design process. Provision had to be made while maintaining correct proportions, to lose excess water from a vessel before it could lead to rhythm arrest.

Two vessel sizes were combined in a more or less symmetrical complex, 30 cm and 40 cm wide on a gentle slope (Fig. 5.11). Adjustments had to be made until the lateral rhythmical swinging movement appeared. But then a whole new spectacle developed. By virtue of the combination of left and right handed circling movements within the cavities a figure-of-eight (lemniscatory) flow-path appeared, now in a quite clear and dramatic way. The stream of water exiting from one vessel to the next, which in the experimental case was the larger one, flowed out of the vessel alternately, moving diagonally to left and right.

As overall vessel dimensions were increased so also the rhythm frequency decreased. Not only did size vary the frequency but interferences brought about a continually changing rhythmical pattern, coinciding with rhythms in the living world where often a longer periodicity sequence is evident. Combinations of different sizes therefore created an in-and-out of step situation, and changing phases became evident between the vessels at one moment supporting and at the next inhibiting the following oscillations (Fig. 5.10). This would prove to be a vital phenomenon in the whole matter of developing rhythms within these types of vessel — which in due course would come to be called Flowforms.[9] (Fig. 5.12)

It was later established that even in cascades composed of a repetition of the same Flowform, depending on the degree of influence allowed to pass between the forms, rhythms of longer periodicity could be established (see Järna Flowform, Fig. 8.2 on p. 113). The method proved with further investigation to enable specific periodicities to be designed for water treatments of different kinds.

During these experiments it was possible already to observe that any material carried in the water, such as sand, collected at two points in the Flowform, round which the left- and right-handed vortices moved. This led to the idea of creating a hole at each of these points, thus allowing the water to exit (Fig. 5.11). Later developments proved this to be an extremely useful method of enhancing the vortical movement (see Vortex Flowform and the Herrmannsdorf project, p. 111).

Figure 5.10 (opposite left). An attempt to show in a flow drawing, changing frequencies of rhythm leading to support or collapse of movement in the following Flowform.

Figure 5.11 (opposite right). A notebook page indicating the experiment which for the first time demonstrated the full lemniscatory movement. Below the idea of making holes in the base at the point where deposition of material is always seen.

Figure 5.12 (left). Naming the discovery. After many thoughts and suggestions, we arrived at the eventual and more professional style for the name 'Flowform' based on my original design for a letterhead and logo.

Figure 5.13. Some sequences of the double ceramic units set up in Herrischried for experimentation during the summer of 1970.

Further experiments

Let us look back again for a moment at what has happened in the story up to now. First of all, observations of the path of vortices were made. The conclusion was drawn that water has a certain potential for order and metamorphic development in its movement.

'Can something be done with this?' was the next question. And following that question: 'What part has symmetry to play in the realm of fluidic asymmetry?'

Experiments revealed the genesis of rhythm, which evidently comes about through resistance, always of a very specific proportion.

It was a revelation to experience how, through paying attention to subtleties of proportion, an Art form can relate to the world of living organisms, revealing that rhythm was the uniting factor. It was also inspiring to experience the wealth of diversity generated by the marriage of water with rhythms and surface.

It seemed, then, that in the elevated realm of living beings, consistency of rhythm was of indisputable importance. Thus I arrived at the question: how, out of the infinite multiplicity or even chaos of rhythms in outer nature, in the mountain stream for instance, could one organize and emphasize the specific to achieve an enhancement in life sustaining capacity? How was this to be accomplished?

Following the early findings at Herrischried during April 1970, any number of ideas and experimentation followed as a natural consequence of this question. Many ideas were turned over in my mind, some investigated practically, some also discussed with Theodor Schwenk. One important decision was made, namely to resume work in the summer with the aim of investigating qualitative influences with the Drop Picture method (see p. 144 and Schwenk's *Bewegungsformen des Wassers*). To this end, specially designed ceramic forms needed to be prepared back in England.

With the help of Annette Lychou, I made a series of simple Flowforms in ceramic. This was a first attempt with very little time to experiment. The unit consisted of an entry cavity followed by four Flowforms which gradually increased in size, made in two pieces to facilitate firing and transport (Fig. 5.13).

Figure 5.14. This picture anticipates the much later developed Malmö Flowform with a special end Flowform and exit lip which accentuates the swinging movement of the waterfall. See p. 119 and Figures 8.17a-c for details.

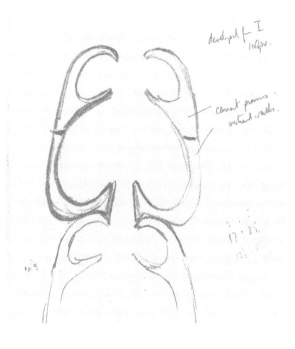

Figure 5.15 (above). Double units with cavities of different sizes could be used.
Figure 5.16 (opposite). A notebook page with some ideas.

The gradient chosen was quite gentle and the forms shallow. Although the rhythmical lemniscatory movements were visible the vigour was minimal. This was a learning process and the resulting forms did not produce dramatic enough movement to register any change in the quality of the water later tested in Stutzhof with the Drop Picture method. Nonetheless, after witnessing the experiment, Theodor Schwenk commented that he observed a change in the quality of the movement over a period of time, from a more passive tired appearance to a more active sparkling one.

What was also observable was that the water discharged from the unit showed a longer periodic rhythm. The several rhythms, generated by the varied sizes of Flowforms, demonstrated a cumulative effect on the final waterfall, so the water flowing through was seen to carry a composite of all the rhythms induced. This is reminiscent of the fluctuating rhythms within the living organism. Indeed the rhythm in the last and largest form of this sequence continually changed under the influence of the preceding ones.

The waterfall rhythms also distinctly showed a swinging left- and right-handed movement resulting from the exiting of water from the lateral cavities of the Flowform as it pendulated from side to side (an example, developed much later, indicates how this can be used to good effect and is shown in Fig. 5.14 and also in Figs 8.17a–c on p. 119).

During three weeks at Stutzhof beginning on July 20th, a number of further trials were carried out.

Flow over a weir

On a flat sloping surface such as a weir, over which water flows in a relatively thin film, I had the idea of dividing up the surface by means of shaped concrete blocks, organized to combine as units creating Flowform cavities through which the water would be brought into multiple rhythmic movements. If such movements were shown to have positive effects then this could be a simple way of achieving them. There would be many ways in which this could be done. Some sketch book ideas are illustrated here (Figs. 5.15–5.16).

Mathematical surfaces

It had already become evident to me in April that the work begun with George Adams (see Appendix 3) on the investigation of mathematical surfaces and their possible effect upon the quality of

FLOWFORMS: THE RHYTHMIC POWER OF WATER

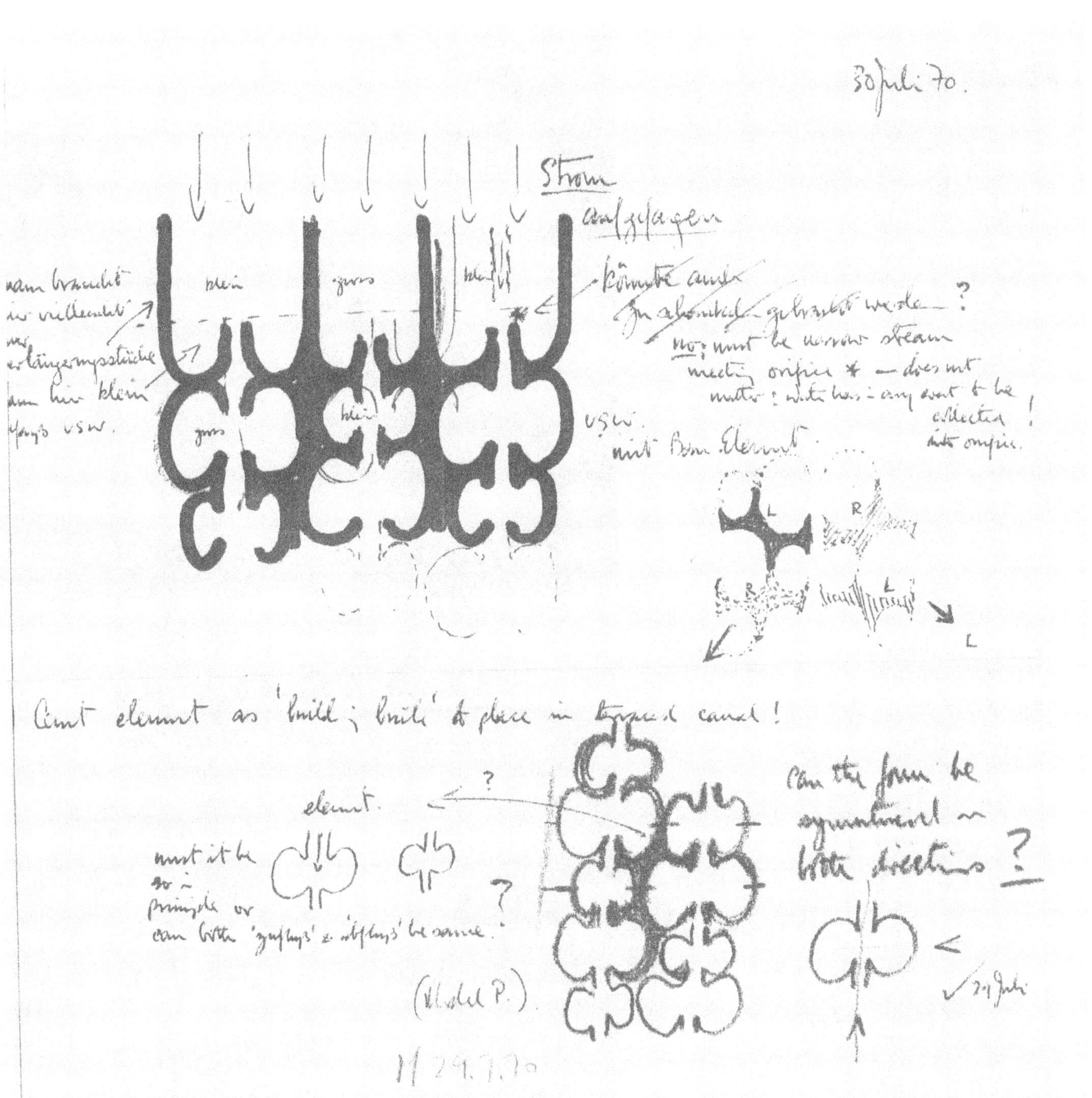
30 Juli 70.
Strom
klein
gross
usw.
element.
Can the form be
both directions ?
(Model P.)

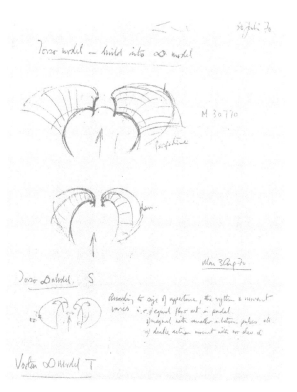

Figure 5.17 (above left). Sketches for the idea of building mathematical path-curve surfaces into the Flowform.

Figures 5.18 (above right). A first demonstration of the inclusion of mathematical path-curve surfaces in a Flowform. This proves an ideal way of encouraging water to spread out in a thin layer over the surfaces in question, thus caressing them intimately.

Figure 5.19 (below). The first indication of a radial Flowform with central inlet and water flowing in three directions. It was eventually developed so that the whole surface would be wet, thus leaving no dry marks. The exits with convex lip create dome-like pulsing waterfalls.

water, could be incorporated into this technique with rhythmical movement.

To demonstrate this it was necessary to take a section of one of the existing surfaces, create a mirror image of it, then bring the left- and right-handed sections together in such a relationship that water passing between them would be brought into an oscillation (Fig. 5.18). The resulting movement would carry the water at best in a thin wave over the laterally orientated surfaces, caressing them intimately.

This experiment was successful, demonstrating that the Flowform Method provided a way forward to achieve a possible optimization effect upon water by relating the rhythmic movements to mathematical surfaces to which living forms themselves have a close affinity.

This was the significant breakthrough that fully justified all the efforts of the 1960s to create the spectrum of mathematical surfaces made possible through the calculations of George Adams. Such surfaces could now be investigated for their qualitative effect upon water, work which continues to be central in the activities of the Virbela Rhythm Research Institute in Sussex (see Appendix 4).

FLOWFORMS: THE RHYTHMIC POWER OF WATER

Radial Flowform

On August 1st, the idea of a radially orientated Flowform was noted (Fig. 5.19). This was first considered for a project in Fallun, Sweden. Later it was worked out as a small-scale model for the three metre diameter Amsterdam Flowform used in the Dutch Floriade of 1982. It made sense to finalize the small form as well and this later became the Ashdown Flowform used first at the Ashdown Health Centre and subsequently for many private gardens (Fig. 5.20).

Tube Flowforms and other ideas

During my next period in Sweden in March 1971, I began working with a very simple system, made by cutting sections of pipe to create Flowforms with a rigorous geometrical character (Fig. 5.21). This form proved itself to be a useful flexible way of examining and describing the parameters of the Flowform Method in simple mathematical terms for patent purposes. Even a four-cavity Flowform became possible (Fig. 5.22).

Other ideas were investigated for the use of Flowforms as step units, straight or spiralling round a central axis, (Fig.5.23) or even

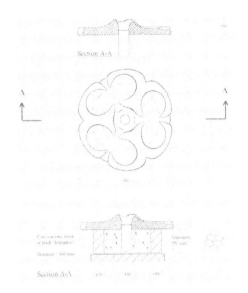

Figure 5.20 (above). The Ashdown model.

Figure 5.21 (below left). Ideas for very simple cylindrical units made from pipe sections. These could be made very simply in large scale by using pre-cast units off the shelf.

Figure 5.22 (below right). A four cavity unit which generates complex movements (see Chp. 11)

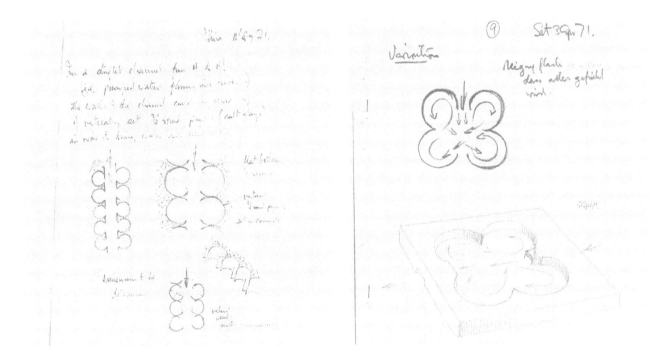

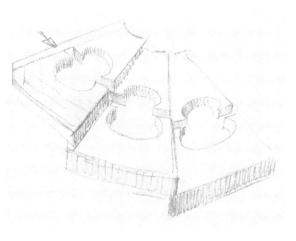

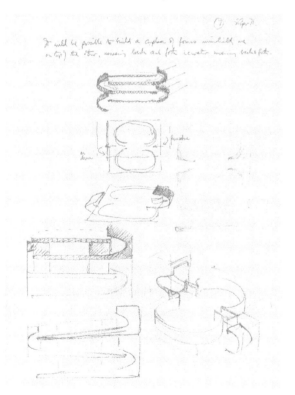

Figures 5.23 (above). Developing the principle of steps in which cavities could be formed as a Flowform cascade.

Figure 5.24 (below). The first ideas for developing stackable units with linking channels.

Figure 5.25 (below right). Ceramic rings for simple construction within existing channels to enliven the water movement, from laminar to harmonic.

for stacking vertically (Fig. 5.24). Ceramic rings were also conceived (Fig. 5.25). See Figure 8.4 on page 114.

A further idea for bringing water back to the entry, letting it exit below or allowing it to move round an inner and outer ring offered other opportunities for investigation (Fig. 5.26).

Water circulation

General questions concerning the circulation of water already occupied me, especially the negative effects of pump pressures and mechanically induced turbulence which are so degenerative of water quality. Through the Flowform method, I wanted to find ways of transporting water, especially for treatments relating to the quality improvement of water, that avoided such influences.

The issue that would occupy me for years to come was that of quality improvement through rhythmical movement. I needed to be able to create a treatment which would be as 'pure' as possible, that is, avoiding the influences of pressure, materials and electromagnetic fields that could cover or negate the effects of rhythm. Perhaps a rocker could be developed that would carry water repeatedly between two reservoirs (Figs. 5.27, 5.28). For this purpose Flowforms would be needed that would function in both directions! My working sketch (Fig. 5.29) became a working model (Fig. 5.30).

FLOWFORMS: THE RHYTHMIC POWER OF WATER

This rocker installation was carried out at Emerson College in Sussex, U.K., during the following months. It was however difficult to achieve a long enough duration of treatment between the two end reservoirs. With such limited capacity, the available water emptied out of the reservoir before movement in the Flowform sequence had really developed properly. More work still needs to be done on dimensioning and capacity.

I also had the idea of placing Flowforms round the perimeter of a large disc, perhaps several metres in diameter, with an elevated pivot in the centre giving a slope to the disc which would then be rotated. This would certainly function for longer treatments and quality tests but, in the end, I did not manage to carry it out. It would need a large number of suitably designed, stoneware Flowforms.

Figure 5.26 (above). Sketch for a Flowform with perimeter channel instead of direct flow-through. This could be useful for a stacking procedure. Exit is below.
Figure 5.27 (below left). Sketches for rocking apparatus to perform long term treatment without pumping.
Figure 5.28 (below right). Construction ideas for a cradle.

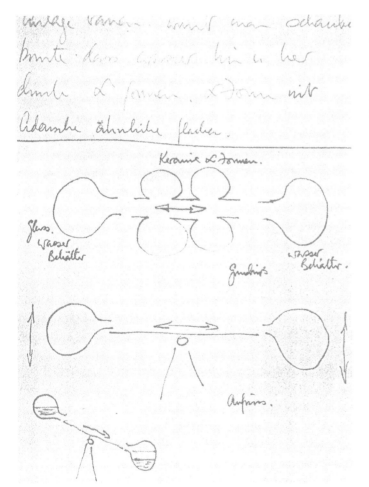

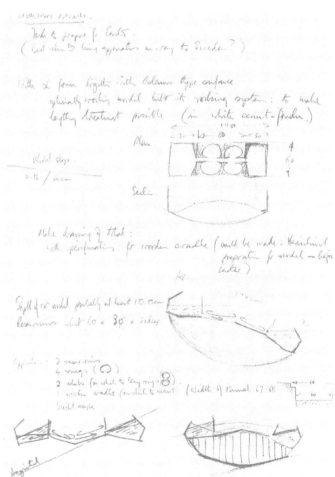

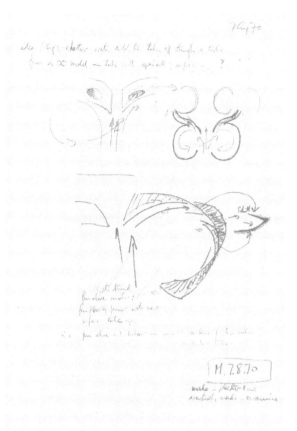

Figure 5.29 (left). A Flowform unit that would function in both directions.

Figure 5.30 (right). The actual units for production in ceramic material, probably using mathematical spiralling surfaces.

As a result of these periods of experimentation it became increasingly clear that there would be numerous ways in which the Flowform Method could be implemented, leading to a whole range of models and applications.

Working from questions

These early practical researches generated a number of questions, both in my own work and, after 1974, with colleagues who joined me. As so much of our progress arises from the simple asking of questions, I offer some of these considerations below, trusting that the reader will be prepared to share in this questioning approach without necessarily expecting comprehensive answers. It is good to live with questions and one can only hope that they generate others.

— As the living organism appears always to exist in some rhythmical context what is the significance of this? Life either exists within daily and seasonal rhythms as with the plant, or generates a rhythm within its own fluid system as with more independent higher organisms.

— Has rhythm some effect upon the fluid processes and substances themselves?

— Could the quality of water brought out of a chaos of rhythms into an ordered spectrum of rhythms, be changed or influenced in a positive way?

— Could rhythmically treated water support plant, animal or human in a more effective way?

— What part does surface play?

— What is quality?

— Can water be something other than H_2O?

— In what way does it mediate qualities beyond its own chemical nature?

— What is water's actual function? Is it merely a fluid base which dissolves minerals?

— What is wetness?

Some factors for consideration

Water obviously supports life. This function is however jeopardized by the addition of certain substances in a given excess, by different grades of warmth and by the way it is moved.

In other words it can become a hazard to life through poisons; through heat; through the vigour of its movement generated by gravity, or pressure forces whether technical or natural.

Unnatural movement under pressure can be destructive to water's subsequent capacity to support life, where no obvious chemical change has taken place. Mountain farmers complain of a deterioration of irrigation water taken below hydroelectric turbines.

Water needs our consideration and support, to heal it after we have used it for technological purposes.

Our technological use of water is unlikely to decrease but our efforts to heal it must increase if we are to survive.

Figure 5.31. What has now been achieved is indicated in this diagram. The sequence of forms can be understood to show a time process during which water passing through a cavity gradually comes into a lemniscatory movement.

Wherever possible our use of water must also be questioned; do we simply continue to 'make water work' as the phrase so often goes, or do we develop an enlightened attitude to water and to its life supporting function.

Repeatedly we have to think of life as rhythm. Rhythm by its very nature presents a wide spectrum of qualities, it is fundamental to metamorphosis, which is boundless in nature.

Healing will hardly be achieved with single technical solutions. We must work with nature's multiplicity; for every situation a unique set of conditions exists. To penetrate this field of form and rhythm, to understand and work with it as a universal force seems the main way open. It is not so much a question of deciding if this is indeed the way to work with water, but rather how can an optimum handling be achieved.

Thoughts such as these were always being formulated and considered as I continued to work. These same considerations continue to guide us even now.

Figure 5.32. Here is indicated diagrammatically the achieving of the lemniscatory movement by virtue of the establishing of correct proportions.

6

The Flowform and the Living World

The Flowform and life processes

In creating an organ for water we can also think of it in terms of seven life-processes described by Rudolf Steiner (1990, lecture of August 12, 1916). Every living organism:

(1) receives substances; which are
(2) accepted; and
(3) digested. The organism responds with
(4) certain secretive processes,
(5) is nourished,
(6) grows; and
(7) eventually reproduces itself.
(See also Table 4 on p. 123.)

Rhythms are the vehicle for the life process. In the human organism, for instance, two strong rhythms among many other more hidden subtle ones are evident in the breathing and blood circulation. These relate with each other in ever minutely fluctuating proportions.

In the case of the Flowform, the physical organ itself is of course totally passive. The intention is that the water is taken into it and is encouraged to relate to the surfaces in an intimate way, so that rhythmical, pulsing movements are generated. These movements may open up the water in its innermost structure to the influences of the environment. Not only gases are absorbed but also more subtle planetary influences which are evident in relation to the Moon for instance. Reference has been made already to this innermost structure, manifest in veil-like forms which are dramatically changed, and also receptive in their movement, to outside events (see pp. 54f). Whatever changes occur within the water, these are carried on into the environment and influence regenerative biological processes.

Rhythmic patterns are therefore evident in every life process and they reveal, within its fluid body, the signature of an individual organism. If this process is interfered with in a disruptive way, the very existence of the organism can be threatened.

Fluid is taken into the organism in order to mediate these rhythms which in turn carry life. Quite apart from the necessity to cleanse, that is the removal and not the addition of foreign substances, the quality of movement is directly related to water's ability to influence and sustain life processes.

We have repeatedly been able to observe through experimentation that organisms respond differently to water which has been moved in different ways (see p. 135, the Warmonderhof project). In the contemporary world at large, it is the case that increasingly water's more subtle capacity to move is degenerating and is in need of healing.

Our aim is to establish on the one hand the influence of rhythms in general and, on the other which specific rhythms will bring about a healing and revitalizing process when applied to given organisms and under what circumstances this must take place. To this end, it is important to understand those aspects of the Flowform which relate to natural processes.

We shall examine more closely those aspects of the earth's organic whole which are external to us, but first look at the complementary process within living organisms such as the heart. Fluid is taken in through an orifice and as it circulates it expands throughout the organism and contracts again as it is discharged. An expansion is followed by a contraction.

The heart and the Flowform

I begin with an image of the human heart which was presented to me some years ago. We must first imagine a heart full of blood at any moment — it is an object about the size of our fist, containing something like 100cc of fluid which, with the body at rest, is passing through perhaps seventy times a minute. In a matter of seconds, this volume of blood finds itself dispersed throughout the body to its farthest extremities, distributed through the organs and capillaries like a 'cloud' to the very bounds of our skin. The total length of these capillaries in a human body has been estimated by some authors as somewhere between 25,000 and 60,000 miles!

It is the simple pulsating of the organ that produces this miracle, a rhythm which most people accept as the mechanical equivalent of a pump. But is this truly the case? Is the heart really constituted, in its form and mechanical strength, as a pump? The reality seems otherwise. Theodor Schwenk has written that:

> It must not be forgotten that the very pulsating motion itself belongs to the nature of flowing liquid. Taking our start from facts like these, there is no need to imagine the activity of the heart to be that of a pump. In the human heart, form and movement are interrelated, uniting space and time in a rhythmical process. The organ, a form in space, is simultaneously a movement in time (*Sensitive Chaos*, p. 91, 1996 edition).

In 1920 Rudolf Steiner commented:

> What is the common belief about the nature of the human heart? It is regarded as a kind of pump, to send the blood into the various organs. There have been intricate mechanical analogies, in explanation of the heart's action — analogies totally at variance with embryology, be it noted!

Blood, he observed, was propelled with its own biological momentum, as can be seen in the embryo, and boosts itself with 'induced' momenta from the heart; '...pressure,' he said, 'does not cause the blood to circulate but is caused by interrupting the circulation.' He continued:

> The most important fact about the heart is that its activity is not a cause but an effect ... All that can be observed in the heart must be looked upon as an effect, not a cause, [which appears] as a mechanical effect to begin with. The only hopeful investigations on these lines, so far, [1920] have been those of Dr Karl Schmidt on 'The Heart Actions and the Curve of the Pulse.' (Schmidt 1892) The content of this article is comparatively small, but it proves that his medical practice had enlightened the author on the fact that the heart in no way resembled the ordinary pump but rather must be considered a dam-like organ. Schmidt compares cardiac

action to that of the hydraulic ram set in motion by the currents. This is the kernel of truth in this work ...
(Steiner 1975)

Later in the 1920's, Dr Martin Mendelsohn, writing in great detail and with compelling logic about the heart as a 'secondary organ' rather than a central 'mechanism' which generates the circulation, observed that the common view of the heart as a pump: '...is so absurd, it is difficult to understand how it has held credence for so long.' (Mendelsohn 1928)

The heart simply does not have the power (as a small 300 gramme organ) to transport at least 8000 litres/day (with the body at rest) and thus maintain, day in day out, what would be equivalent to lifting 450 kg to a height of 1600 metres. One also has to remember that blood is five times more viscous than water and moves through millions of capillaries which are often finer than the dimensions of the blood corpuscles themselves.

The aorta responds to the movement of the heart in an opposite way to what is expected. We can visualize this by analogy with the Bourdon pressure gauge, a curved tube through which liquid is passed under pressure such that the greater the pressure, the straighter the tube. In the heart, however, during the systole phase when the blood moves on, the aorta has been observed to *increase* its curvature. Researchers have measured a peak flow that markedly precedes peak pressure, which means there is a momentary negative pressure at the exit to the heart. If the blood were being *pushed* out of the heart, the curvature of the aorta would decrease. Any pressure reading is due to interrupted or restricted momentum (see Marinelli 1996.)

Normal pressure in the left ventricle would be likely to cause a rupture of the heart wall at the apex, which is very thin indeed. We need only examine the heart muscle carefully to establish the fact that at the apex the wall is fractionally more than 2 mm thick while the ventrical muscle can be 40 mm thick. This is contrary to the usual drawings in modern textbooks. As a matter of interest I did dissect a bovine heart to see what the apex wall looks like (Fig. 6.1). It is as little as 4 mm thick, in contrast to the ventricle muscle walls of 60 mm or more. We find the same for a sheep heart on a smaller scale.

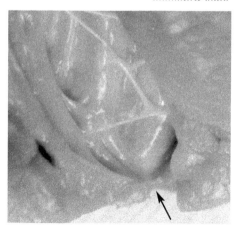

Figure 6.1. The dissection of a cow's heart shows the wall thickness at the apex. Even with the size of this heart the flesh is barely four millimetres thick.

In his 1996 article, R. Marinelli quotes an earlier observation made by J. Bremner of Harvard in 1932:

> He filmed the blood in the very early embryo circulating in self-propelled mode in spiralling streams before the heart was functioning, [and] ... failed to realize that the phenomenon before him had demolished the pressure propulsion principle.

It is also clear, from a physics point of view, that the continually decreasing dimensions of the blood vessels from aorta to capillary, constitute an efficient braking system. When watching fluid movements within tiny vessels one is impressed simply by the vigour and complexity of the phenomenon which in no way appears to be generated from a distant source.

In a recent television programme, a demonstration was given of blood being pumped through a dead heart which had been removed and severed entirely from the body. The heart continued to pulsate, not due to its own pumping action but because it was activated by the stream of blood pumped through it. The organ is thus capable of regulating the stream of fluid passing through by virtue of its resistance to this stream. The resulting pulsatile motion can last many hours after removal from the body, though this may for a while be mutually associated with residual nervous activity. The phenomenon needs further observation.

Rhythm in the heart and in the Flowform

We have seen that when water flows through an aperture or constriction which at the same time offers some resistance, a state of hesitation or instability may occur, and a pulse ensues. This can be demonstrated in the laboratory, using the flexible surface tension of the 'pulsing puddle,' where a stream is offered the right conditions and proportions (see Chapter 4).

In an organism, the most rudimentary vessel through which fluids pass has the general shape of a meander, but this is so proportioned that a pulse is supported. Living organs such as the heart are three-dimensional, closed and flexible, and fluid passing through them is encouraged to pulsate. Once triggered, the flexible or muscular build of the organ then supports the tendency to rhythmical movement, by means of intermittent contractions. In this way suction and pressure, periphery and centre, mutually maintain the

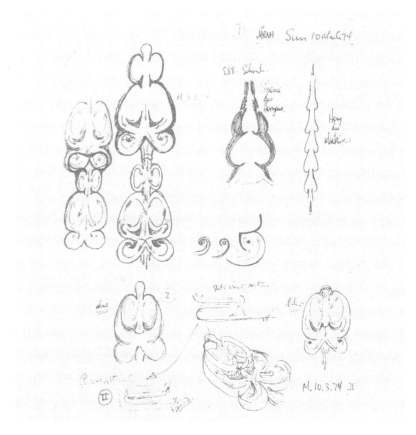

Figure 6.2. A sketch book page on ideas for working models of organ-like Flowforms.

process of regulation and movement. To prevent a stream moving in both directions, the narrow connection between any two cavities is formed as a valve which closes against back-flow.

A cross-section of such an organ is very similar in principle to the plan of a Flowform which provides a rigid, basically open container, allowing water flowing through to maintain a free surface. Living inner organs are however normally flexible (Fig. 6.2).

The typical Flowform sequence in plan is found to be reminiscent of the configuration of heart or digestive organs in some insects. Such organs in, say, a beetle, exhibit a series of cavities more or less symmetrical (Fig. 6.3). The following table offers some comparative ideas on the relationship of the primitive heart of a beetle and a series of Flowforms:

> The way in which the surrounding form, whether flexible or rigid, supports rhythms determines the direction of flow. In the rigid Flowform, the flow of water, generated externally,

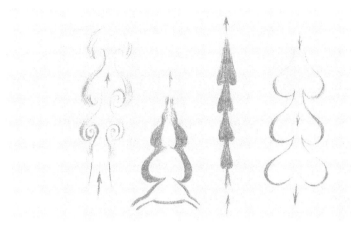

Figure 6.3. These four drawings from left to right show firstly the forms water can generate as it flows out from an orifice under water, a schematic view of a heart, the heart of a beetle (all after illustrations in Sensitive Chaos*) and finally a schematic plan view of the Flowform. It is important to note the direction of flow in each example.*

Primitive heart of a beetle	Flowform Series
Formative processes *per se* direct the creation of organs within the fluids.	In a similar way the designer creates the form of the vessel in conjunction with water movements.
The activity of the fluid is always innately vortical throughout its volume.	Water movement has the quality of manifold convoluting veil-like surfaces within its volume.
Vessels are created in terms of the activity of the fluids.	Designed vessel surfaces have to relate intimately to the characteristics of the moving water.
Flexible membranes provide a three-dimensional enclosed cavity.	Inflexible vessels provide an open, generally horizontal 'two-dimensional' channel.
Flow-through generated by suction or biological activity from the periphery (levity).	Flow-through generated by the slope of the vessel sequence (influence of gravity).
Proportions of the flexible cavity generate systole/diastole regulated flow.	Proportions of cavity generate rhythmical regulated flow dependent upon the freely fluctuating (flexible) surface of the water.
Suitable resistance of the vessel's form generates regulatory rhythm.	Suitable resistance provided by the forward aperture generates rhythmic oscillation.
Due to the resulting contraction and expansion of the organ, valves are needed to guarantee direction of flow.	Slope and the effect of gravity guarantee the direction of flow (thus no valves are needed).

Table 3: Comparative ideas: the Flowform and the beetle heart.

encounters a narrow aperture which creates the resistance from which the oscillations develop. The form is static and the flow is encouraged by a slope, with gradients between successive Flowform vessels. Pulsation is supported by the flexibility of the water's own free surface.

By contrast, in the living organ the flow is generated by suction from the periphery and by the overall biological activity, and regulated by the proportions of the organic form, which is flexible. A pulse is generated to which the muscular organ responds and provides support. In order to maintain the direction of the closed-in flow, valves are needed.

In section, both flexible organ and fixed vessel are similar. Pulsation is the product in both, but the direction of flow is opposite in relation to the general shape. This contrast of flexibility and rigidity relating to direction of flow is fascinating. The organ itself contracts through a nerve stimulus in support of the flow, triggered with a slight delay perhaps by the already generated rhythm. In the Flowform the corresponding episode appears to be the brief increase in gradient between each successive vessel, which lends impetus to the flow. Again, just as the valves of the flexible organ prevent back-flow, so the waterfall and turbulence in the passage between Flowform vessels prevent 'upward' or reverse flow.

The heart appears to regulate the flow of blood while at the same time acting as a sense organ by means of which the condition of the whole organism can be judged. Probably this process also helps in turn to maintain the blood's sensitivity to the functions of the body.

The Flowform and the water cycle

The natural water cycle itself can be considered as a 'taking in', 'utilization or digestion' and a 'passing on.' An overall picture of expansion and contraction is revealed. As described in Chapter 2, we find the following sequence.

From the water vapour in the atmosphere, clouds condense, rain falls and a confluence of tributaries slowly contributes to the growing meandering river which itself can be considered the regulative 'heart' of the system. When the flatness of the coast is reached the water spreads out into the arms of the delta and expands into the ocean from which it again evaporates. The threefold nature of the process manifests at every stage in all conditions of condensation and evaporation while movement mediates. Water is the element of movement *per se*.

Let us first remind ourselves of the form of a river, which is never naturally straight. Out of the very nature of water's innate movement capacities and the resistance of the ground's surface, the meandering process is generated. This meandering form in turn supports a regulative function for the flow of the river and the surrounding ground water. It is as though the multitude of rhythms in a mountain stream are extracted from this water's potential, to appear in the more subdued rhythmical curving of its bed in the plain. Water's left- and right-handed spiralling movements constantly model themselves on this organ of the water cycle which is at its very heart.

The 'heart' function of water in the wider landscape depends upon the healthy state of the river which in turn indicates a great deal about the condition of the total surrounding environment.

It is becoming increasingly evident — through the work of many investigators — that we must think of the watery body of the earth as maintaining a mediating function which conveys the 'information' of the total environment to all living things. It is this embedding of everything living into the totality that is of paramount importance to the continued existence of life on the planet.

Movement and vitality

Consequences follow for living organisms from the character of movement of the water with which they are associated. It is not just incidental that water flows as it does in mountain streambeds, with all the turbulence, rhythm, light and air surrounding it.

When these natural processes and movements are removed, there is a consequent loss of life-supporting quality. For instance, as we saw earlier, farmers downstream from modern hydro-electric systems experience that harvests achieved for generations are suddenly reduced, as the water taken for irrigation after flowing through such systems is not as supportive to plant growth. Water, they say must flow down the river some kilometres before it adequately supports growth processes again (Fig. 6.4). The conducting of water through narrow tubing, subject to rapid gravity fall, appears to bring about a significant loss in its vitality.

At the end of the 1980s, I was present at conversations between farmers and a leading hydro-electric representative in New Zealand

Figure 6.4. Again we anticipate the Malmö Flowform (p. 117) here installed by a mountain stream at Sundet, Mösvatn, Norway. Water from the stream higher up is directed through the vessels which generate an ordered rhythmic swinging movement, like a living pulse, (which this photo illustrates very well) in contrast to the multitude of rhythmic movements in the streambed. If mounted below a hydroelectric turbine at a much larger scale, cast in situ, these more life-related rhythms could improve the vital qualities of the water for irrigation, in a shorter distance.

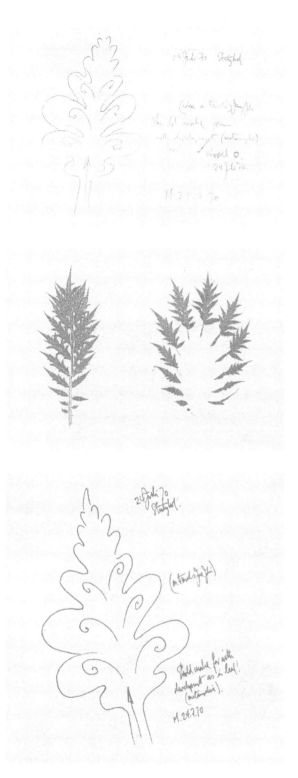

about the possibility, even necessity, of improving water discharge points not only aesthetically but also functionally. Could the water which had been subjected to the constraints of mountainside conduits and excessive influence of gravity be in some way reinstated and regenerated? The agriculturalists' view was that some such regeneration should be offered to the water after its projection through the system.

Research shows that through their natural movement and by virtue of their total ecology, rivers will self-regenerate and purify polluted water over a given distance, dependent of course upon the degree of pollution (see Peter 1994). It may be asked whether this process could be accelerated by the introduction of rhythmical activity which would enhance the river's natural regenerative action (Fig. 6.4, p. 89).

The work of Dr Wolfgang Ludwig of Horb, near Tübingen in the Black Forest, as reported in *Umweltmedizin*, reveals that although public drinking water can be chemically pure, it carries electromagnetic information related to the removed pollutants and can still be harmful. He underlines the fact that vortices are necessary for the preparation of regenerated life-supporting water, a function that in normal circumstances is carried out through nature's hydrological cycle. The purer the water is chemically, Ludwig says, the more effective and durable will repeated vortical processes be in neutralizing the unacceptable electromagnetic frequencies. This is the simplest way to deal with the problem, and it is a task that Flowforms can take over with very positive effect and without necessarily any additional local use of energy.[10]

Such questions are not merely academic. The issue of the vitality of water becomes more and more urgent as the world faces food shortages resulting from over-exploitation of water resources (see Chapter 13).

Developing a metamorphic sequence

Could we create an 'organ' which would enable water to manifest its potential for order and metamorphic change?

This question arose some time before the emergence of the Flowform itself. It was sparked by the observation of the path of vortices, and was framed as a response to the prevailing attitude of 'making water work' where water is almost exclusively considered as a transporting or energy producing medium.

These technological applications entirely ignore the infinitely subtle veil-like forms that fill a body of unadulterated water with harmonic movement. Water's ability to move in a rich vortical manner is related to its capacity for supporting living processes. In our work, therefore, we are always looking at the idea of 're-educating' water to move in an harmonic way, thus nurturing its ability to support life. By encouraging it to move through a spectrum of rhythms and over a rich palette of surface qualities, water might be elevated again towards a nurturing of nature's wisdom-filled formative processes.

Early ideas for cascades

On July 24th, 1970, I began to work with an idea that had come to me during my travels, of a series of Flowforms at first increasing and then decreasing in size like the form of a composite leaf (Fig. 6.5). This was to become my central theme, namely a sevenfold metamorphic sequence, or cascade, offering a whole spectrum of rhythms and gestures to water flowing through, related to the way in which nature builds organs for fluid processes.

The artichoke is one example in nature of such a leaf (Fig. 6.6). Its small lower rounded forms expand towards the middle part of the leaf and stretch out becoming pointed at the end. Such a single leaf anticipates the leaf development process of the whole plant (see J. Bockemühl 1967). A specimen that I prepared shows the complete leaf on the left, while on the right, one side is arranged to show its similarity with a normal leaf development in the whole plant. This is followed by the original 'leaf' sketch leading to the idea of the sevenfold sequence (Fig. 6.7).

Nature's patterns so often repeat themselves, and the fish skeleton in the vertical position reminds us of the pattern of a composite leaf with the largest extension of the forms in the middle region (Fig. 6.8). We can see how the angle of the lateral bone processes sweeps round, from one end to the other, demonstrating the motif eventually built into the Flowform's changing lemniscatory axes. From the leaf, via the fish as transition to the mammal spine, we experience the gradual appearance and definition of the sevenfold (Fig. 6.9, and see p. 60). With the rib-cage included there is also from one end to the other an increased volume in the middle region.

This sequence is derived metamorphically from the spiritual

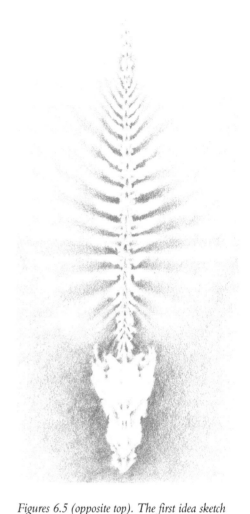

Figures 6.5 (opposite top). The first idea sketch based on the form of a leaf, for a Flowform cascade of seven metamorphic stages. 24 July 1970.

Figure 6.6 (opposite middle). The whole artichoke leaf also demonstrating the swinging movement of its composite elements from the base to the point. On the right, one side of the whole leaf, showing its relationship to the metamorphic development in the whole plant. (It is unfortunately not possible to show an example of the whole plant as it is very large and such a preparation was not undertaken by the author!)

Figure 6.7 (opposite below). Original sketch which led finally to the Sevenfold Cascade idea.

Figure 6.8 (above). A fish spine shown as a transition between a composite leaf and a mammal spine. Drawn by Helen Aurell.

Figure 6.9. The spine of an herbivore (Llama) from which one can differentiate seven phases in the groups of bones, within the cranium, atlas/axis, cervical, dorsal, lumber, sacral and caudal. Drawn by Helen Aurell.

archetype of the human form, the 'Idea' from which all the organic aspects of nature are derived, mineral, plant and animal. Each organism manifests a facet of the archetype by virtue of its physical-mineral body, the life body and the realm of consciousness.

The ribs themselves also demonstrate the gradual sweep from the front to the back, or, for instance in the human skeleton, from the horizontal in the upper dorsal region, to vertical in the lower dorsal region.

The 'cranial' bones of the spine within the skull, come to an end or 'disappear' in expansive or planar processes (the vomer ends as a delicate channel most clearly shown in the animal snout), at the tail end the spine disappears in contractive or linear processes (limb-like bones). At opposite ends a certain simplicity is demonstrated but of polar quality, while in the middle is the complex realm of rhythmical interrelationships. From end to end, one can describe a threefold entry and a threefold exit with a mediating central realm.

The 'ideal' Flowform

In the initial three stages of a cascade one can consider three qualities of movement: polar opposites of fast and slow, with an harmonious combination in between. We can ask: is there an ideal Flowform, of specific size and shape which will provide an optimal quality of rhythm for water with its particular viscosity? This would lie between these two opposite tendencies which generate on the one hand a nervous, fast movement in a smaller ves-

sel and on the other a more ponderous slow movement in the larger vessel. This last one would in turn be the one which 'digests' all the preceding rhythms. The optimal central form would tend to determine the size of the whole sequence. With the last three stages one could consider the same questions. And finally with the whole sevenfold series. The first section, the last section and the middle Flowform as balancing entity.

An answer to the question regarding an 'ideal' Flowform may be that there would be a range, the relative proportions of which would be of major significance. The enhancement effect would also be part of the digesting process indicated.

These permutations are complex but through careful design of the Flowforms they can be quite specifically chosen. A moving pattern of more or less pre-planned rhythms can appear. It seems feasible to assume that this is similar to what happens in any living organism. An existing combination of rhythms and forms, catalysed by the seed is instrumental in building an organism which then itself behaves as a mediator for these rhythms — especially through its fluid organization.

To summarize, such similarities as we have noted may be seen as confirming the idea that with the Flowform we correspondingly create an 'organ for water.' Rhythm regulates the flow while at the same time it re-sensitizes the water, increasing its potency of mediation. The underlying theme of a resulting qualitative change represents the central issue of this study.

The Flowform method demonstrates, in a totally different context outside the organism, the fact that rhythmical processes are generated through resistance of a very specific order in a given context of flow.

In the following section, we look at applications of the Flowform and, in Chapter 9, examine in more detail the different forms of the sevenfold cascade that have been achieved.

PART 3

Applications and Research

7

Järna: the first major Flowform project

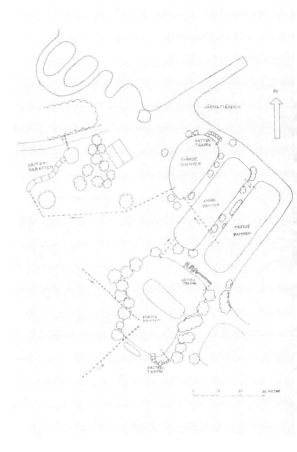

Figure 7.1. General plan of the biological sewage system in Järna as it developed after a few years.

During the summer of 1973 the first opportunity arose through the invitation of the well-known painter, designer and educator Arne Klingborg, to design and build a major Flowform project. The site was the Rudolf Steiner Adult Education College founded by him in the early sixties in Järna, on the Baltic Sea fifty kilometres south of Stockholm. Small biological treatment ponds had been in use there for some time and were in need of enlargement to accommodate an increase in population on the site. The new system was required to handle the total sewage from a community of up to two hundred people.

Klingborg did not favour the 'out of sight out of mind' approach for dealing with waste products. His idea was to create a beautiful water garden, to provide an amenity and at the same time a solution to the problem of dirty water which must be prepared for re-entry into the natural cycle, in this event into the sea.

On a visit to Emerson College in June 1973, Arne Klingborg came to my studio to see what had been developing since 1970. His vision of creating a water garden was to allow for the demonstration of a number of aspects of the natural biological cycles. Flowforms, he thought, would provide a suitable cascading system to introduce movement into an otherwise very static landscape. With Flowforms, rhythms would be introduced into the body of water, creating an environment also conducive to oxygen enhancement in support of aerobic processes.

In the domestic life of the community at Järna, every attempt is made to reduce and even eliminate the use of toxic substances which do not conform to an acceptable degree of biodegradability. These are in any event damaging to nature. The surrounding gardens and park, the vegetable plots of the college and the market gardens nearby, as well as many acres of surrounding farmland are

all run biodynamically with great success. This method of organic farming precludes the use of any form of chemical additives. A plan of the water garden area is shown in Fig. 7.1.

Our work with Flowforms here entered a new phase. I had posed the question earlier whether Flowforms could improve the condition of effluent issuing as discharge from a sewage plant, in order to aid its re-entry into the natural cycle. Through the Järna project they were brought into the biological purification and regenerative processes as such.

The Flowform contribution we were considering had primarily to do with the effects of rhythm on the biological system, not simply the extent to which Flowforms increase oxygenation. In the natural regenerative processes it is the organisms which extract and digest substances from the water and these are all strongly rhythmical.

Watching the activities of micro-organisms through a microscope for a few moments, is very revealing. They are in continual rhythmical movement. They are so at one with their environment that one might say they 'are' their own environment. They constitute an intimate aspect of a totality and move with it, always in rhythms, nothing ever being repeated. The planets move out there in space, and this hierarchy of organisms all move in response to them. It seems we have to penetrate the significance of this multitude of movements.

The world has thrived on micro-organisms for millennia. Now our task is to understand and not merely to disrupt and destroy them. It is this perspective which I would ask the reader to keep in mind as an underlying theme of this work with water and environment.

It was decided to dedicate some weeks during that summer to designing and building the cascades at Järna. I took one small preliminary Flowform design with me (Fig. 7.2) to Sweden and made a further two designs of different sizes (Figs. 7.3, 7.4). The interior surfaces were rubber-moulded and casts made in glass-reinforced cement, such an unsophisticated method being employed due to the necessity for speed of production.[11]

On a mound of newly excavated material we were able to use a discarded iron girder as foundation for some twenty of the small Flowforms. This necessitated a straight plan (Fig. 7.5).

The two original ponds had been made into one, leaving an island in the middle with a row of trees. On this island Lars Fredlund, an architect and teacher involved in the development of the area, mounted a long log on a trestle supported by the trees. This held

FLOWFORMS: THE RHYTHMIC POWER OF WATER

Figure 7.2 (opposite top). (010670) In 1973 the first cascade with basic Flowform I took from England to start off the process. All the Flowforms in this summer were made in cement Fondu (a quick setting alumina cement) using fibre-glass and sand for the shell casts made from rubber moulds.

Figure 7.3 (opposite). (170773) Prototype for the Slurry Flowform made in Järna and moulded for casting.

Figure 7.4 (opposite). (270773) Prototype of the Sewage Flowform also made in Järna as modification and slight enlargement of the Slurry.

Figure 7.5 (above left). In the next summer the planting was becoming established.

Figure 7.6 (above). The beam set up by Lars Fredlund for the random sequence of Basic, Slurry and Sewage Flowforms

Figure 7.7 (below). A close-up of this beam cascade functioning.

twenty Flowforms of the three different sizes available (Fig. 7.6). The main emphasis was on the rhythmical movements (Fig. 7.7).

Although the installation was not designed for comparative research (see p. 135 on the installation at Warmonderhof) it has been continually developed in the process of improving its efficacy. Lars Fredlund monitored it for many years and the authorities have made regular tests. Now, twenty-seven and more years after it was built and other people have become involved, a major cycle has passed and extensions and changes are being made. In the meantime initiatives to further this kind of system have developed in many countries.

With the agreement of Pele Larson of Ohlson and Skarne, we were able to use the extra Akalla Flowform casts we made at their Märsta factory and build a new cascade in Järna during the summer of 1976. This work was carried out by Nigel Wells, Paul Farrell and Niels Sonne Fredrikson (Fig. 7.8 to 7.11). The cascade shows a random metamorphosis of three sizes of Flowform which function with the same flow-rate, the small ones generating a vigorous fast

Figure 7.8 (top left) Setting up on columns the Akalla Flowforms in Järna, carried out by Nigel Wells, Paul Farrell and Niels Sonne-Fredriksen.

Figure 7.9 (top right) and Figure 7.10 (lower left) Casting further large Akalla Flowforms. This work had to be carried out without the conveniences of the factory in Märsta.

Figure 7.10 (lower right) Paul Farrell resetting the Järna Flowform cascade.

rhythm compared with the large two-metre diameter Flowforms (for details see Figs. 7.13 to 7.17).

During the summer Paul Farrell also rebuilt the Järna Flowform cascade using the design revised by Nigel Wells.

Earlier, in February 1976 the Waldorf Education exhibition was taken to Malmö in south Sweden and for this another larger cascade had been built. A new design was used which became the Malmö Flowform. In April 1976 this was shown at the Stockholm Modern Art Museum in the Ararat exhibition. Finally these Flowforms were permanently mounted next to the fourth pond in Järna.

Later, during 1979, I conceived the idea of a demonstration channel part of which was built in England by Mark Baxter. It was transported to Järna in November, cast and set up for the major exhibition of 1980. It was temporarily built into the sewage system

after the fourth pond. This was in front of the meander-plant-filter-bed that was now added for final treatment, as well as for study purposes. A small collecting pond issued into the final fish pond on the coast which discharged into the natural reed beds in the sea.

In this way, stage by stage over the years, the treatment path through the ponds, cascades and filter beds was extended.

As the Flowform cascades function the year round, the question of freezing arises (Fig. 7.13–14). Normally ice builds up from the edges and gradually encloses the water, but as it is moving constantly this function is not inhibited. A spacious cavity is created on top on which snow can accumulate, which in turn insulates the whole process. Blow holes usually remain which guarantee adequate air exchange. Quite apart from the function, the most fascinating and beautiful ice formations (Fig 7.17) can be observed. Also acoustically the phenomenon is of interest.

With the temperature rise in the spring, biological activity

Figures 7.12–16 (top to bottom right). With the good wishes of Pele Larson of Ohlsen and Skarne we were able to take the extra unused Akalla casts to Järna as the basis for a further major cascade in the purification pond system.

Figure 7.13–14. Ice building up with a dusting of snow and thawing showing momentarily a beautiful ice spiral in mid-air.

Figure 7.15–16. The small and the large Akalla in spring.

Figure 7.17 (below). The Järna Flowform cascade (Fig. 7.10) 25 years later in winter with accumulating ice.

CHAPTER 7. JÄRNA: THE FIRST MAJOR FLOWFORM PROJECT

Vattentrappa i dammanläggning

increases. Indeed due to the substantial oxygen exchange during the cold period, this activity starts up immediately. Without the cascades it took some weeks. The extra winter storage of liquid effluent is gradually reduced, the sludge build-up is digested and an overall annual balance is maintained. But despite this general tendency, after many years sludge build-up has taken place and further amendments have been made. The main principle is now to separate the solids at the beginning and treat the resulting sludge and liquid separately. Sludge can be detoxified and dehydrated very effectively in specially built and drained reed-beds. The cycle lasts for a period of ten years before emptying, after which a a new cycle can start with the same plants.

Various plant species are chosen for specific purposes and during the summer create a profusion of colour (Fig. 7.19). *Petasites Alba* for example does well round the first pond and is known for its ability to absorb soapy materials. *Scirpus Lacustris*, the common bulrush, round the second pond takes up mineral salts and heavy metals through its roots while various species of willow are used round the fourth pond.

An interesting summary of the functioning of the Järna installation has been provided by the environmental writer Peter Bunyard:

One fundamental difference between the Järna ponds and a conventional sewage works is that whereas the latter attempts to maintain a standard ecology with the production of a purified effluent, at the Seminariet the ecology of the untreated effluent varies dramatically with the changing seasons. In winter when temperatures fall to minus 15°C and lower, little biological activity takes place and solid matter spreads across the bottom of the large pond as a heavy sludge. Despite the coating of ice the cascades continue running and because of the low temperature, are carrying a greater oxygen load than in the summer when solubility of gases decreases. The bacterial count increases substantially over the winter months, purification is slowed down significantly so storage is increased and less effluent discharged. The coliform bacteria, sensitive to 35°C reaches a concentration of nearly half a million per litre and the thermostable coliform bacteria a concentration of 230,000 per litre. By April the bacterial count is negligible, down to 49 and 8 per litre respectively. The situation then remains stable until November when the bacterial count begins to rise again. Algal growth begins surprisingly early in the year, with Chlamydomonas colonizing the surface areas of the ponds in January even though they are still covered with ice. A great many other algae and protozoa follow, Euglena, Cryptomonas, Phacus, Spondylomorum and various paramecium-like organisms. As the water proceeds from one pond to the next the overall ecology seems to change, thus substantiating Klingborg's idea of converting sewage waste into higher forms. Whereas ducks, moorhens, frogs and fish take happily to the third and forth ponds, they are rarely found in the first and second. Working together with Klingborg, Fredlund has planted an interesting selection of vegetation around the edges of the ponds, and clearly the root systems are taking up the minerals, thus contributing to the overall ecological cycle (Bunyard 1977).

Figure 7.19. A summer picture of the beautiful gardens flourishing round the cascade, planted by Arne Klingborg and his helpers.

As Bunyard states, it is not so much a matter of just removing pollutants but of transforming them into higher forms. The water is running through stable forms of organic activity so that minerals are taken up at all levels of complexity. The challenge is to establish an ecology whereby phosphorous for instance passes stepwise up the living hierarchical system until incorporated into higher forms. Thus when phosphorus passes out into the Baltic, it should be in a state that can be assimilated by fish and help build their skeletons.

Further sewage treatment installations

After the first project of this type was built in Järna, from 1973 onwards there followed a number of other installations, some of which are illustrated in a list of Flowform biological sewage plants on p.106. A major project, which was to become also an important centre of comparative research, was created at the Warmonderhof Farm School in Holland. A fuller account of this important project is given in Chapter 10.

Directly as a result of the Järna project, colleagues in Norway — on the initiative of Lars Henrik Nessheim, a leading member of the Camphill Community movement — began building installations at the community villages in Vidaråsen, Solborg-Alm near Oslo, Jösåssen and later Vallesund near Trondheim, and Hogganvik near Vikkedal on the west coast. This initiative was taken up in England by Uwe Burka at the Camphill Community, Oaklands Park, Gloucestershire, and led to the founding of Camphill Water and extended activity across Europe.

Usually these installations were intended as rhythm-enhancing cascades to support the predominantly rhythmic biological processes which digest the toxic and other pollutants in the water. It was never intended or claimed that oxygenation was the main function of the Flowform. Naturally however oxygen content is enhanced wherever vigorous movement is introduced. It has been suggested that oxygen content is stabilized satisfactorily after the use of Flowforms where the idea is to treat water in such a way that it draws in oxygen rather than have air mechanically forced through, which is reportedly only 5% effective. The aim with Flowforms is to circulate in such a way that damage is not inflicted on the micro-organisms one is endeavouring to support, but that a supportive rhythmical environment is created.

These installations have worked very well, and have been looked upon with much interest by the local authorities in Norway.[12]

Treatment of organic matter

Flowforms have been used in connection with tanks containing cow or pig slurry, for example on Murtle Farm, Beildside, Aberdeen. The slurry is pumped from the tanks through Flowforms which oxygenate and vitalize the slurry before it is sprayed on to fields.

As a result of a 1970s research project by Mark Baxter at Tablehurst Farm, Forest Row, Sussex, it was found that treatment in a Flowform reduces the slurry's typical strong smell significantly and provides an improved fertilizer.

Related botanical research

In the late seventies I came across the research of Professor Käthe Seidel of the Limnological Research Institute in Krefeld, Germany, who had worked for over thirty years with reed-bed and other plant systems for water purification. After completing a doctoral study on the common reed, *Phragmites Communalis*, she went on to investigate several thousand plants for their capacity to extract pollutants from water bodies while at the same time withstanding their toxicity. With colleagues, Professor Seidel developed systems for water purification in which changing ecosystems develop and, over a number of stages, different plants such as *Phragmites*, the False Bulrush (*Lacustris Schoenaplectus*) and Yellow Flag (*Iris pseudocorus*) are used.

It became increasingly clear that where lagoon systems could be employed, and where enough space was available, reed-beds would also prove useful for supplementing treatments. Reed-beds were added for instance in the existing Järna installation with good effect.

The possibility of combining all or some of these systems in one complex is proving increasingly interesting: pre-screening, septic tank, settling ponds, sequence of filter beds, fish ponds, with Flowform cascades at various stages.

During the 1970s in the U.S.A, Biological Water Purification Inc., a licensee working with the Seidel system, developed reed-beds for sludge dehydration and detoxification, and for the production of humus over a ten-year bed utilization period. This is an extremely useful method in which solids are separated out and loaded periodically on to the reed-beds. Excess liquid is drained out and returned to the water treatment reed-beds.

According to municipal test reports, after this prolonged composting the product is indistinguishable from good humus.

A recent initiative for biological treatment of local sewage using Flowforms was started in 1998 by Vimala Achuthan in Bangalore, India. To date there are very few or no municipal sewage systems available generally in India, and if this project finds support the use of such systems would be highly appropriate in this large yet densely populated country. We hope to have further contact to maintain the connection for the future.

Biological sewage plants using Flowforms

Järna Rudolf Steiner Högskolan (Project 01 Sweden)
This was the first major project to be carried out with Flowforms and is described in detail in Chapter 7. It began in the summer of 1973. The original Järna cascade still survives, the Akalla, Malmö and Sevenfold cascades have been added.

Warmonderhof State Training in Biodynamics, Tiel (Project 56 Netherlands)
Planned from 1976 with Professor Jan Diek van Mansfeld and built during 1978. A research project with two separate biological systems for comparative work with a Step cascade and three Flowform cascades, Järna, Olympia and Malmö. Described in more detail in Chapter 10.

Vidaråsen (Project 122 Norway)

Built in 1977 it is one of the earlier projects constructed after the first one in Järna, Sweden. During 1998 it was renovated by Prof. Petter Jensen of Aas University, an international consultant in this area of work, and in addition to the original Malmö cascade a further Vortex Flowform cascade was added.

Figure 7.20. Düsseldorf Waldorf School with the rain water retention pond. This water is circulated through a plant biotope system made up of a meandering filterbed and cascade. It is to be used as an educational tool for the study of plants and macro-fauna etc.

Solborg-Alm (Project 206 Norway)

This installation in Norway has been in operation since 1978. An investigation was carried out during 1990–91 by Prof. Petter Jensen and colleagues and a report published (see note 11). It consists of two ponds with Malmö and Järna cascades.

Düsseldorf Waldorfschule (Project 306 Germany)

This school project of the early 1980s from Asmussen's Architect Office in Järna included a holding pond for rain water. This water is treated by means of a meandering filter-bed system and Akalla Flowform cascade before being discharged into the ground. The whole system is intended as an educational tool for use in curriculum work (Fig. 7.20).

Ökosiedlung Gänsersdorf (Project 564 Austria)

A project carried out by Atelier Dreiseitl in 1984 near Vienna within a community housing project. It consists of reedbeds and the Malmö Flowform cascade.

Jössåsen (Project 578 Norway)

An installation built during 1984 in a very out of the way country area east of Trondheim. Several cascades using Järna and Emerson Flowforms were incorporated in conjunction with ponds.

Oaklands Park (Project 630 U.K.)

In 1985 Uwe Burka built a single house cleaning system adding a Sevenfold I cascade in 1987 (Project 725 U.K.). A second much larger system was completed in 1990 with Järna Flowforms and a Sevenfold II Cascade. Burka carried out a great deal of fundamental research and gradually built up an expertise which led to the founding of Camphill Water and a number of installations across Europe. He now functions as a consultant in Weimar, Germany concerned with many aspects of community affairs.

Hogganvik (Project 759 Norway)

A project carried out by Iris Water, on the west coast of Norway during 1988. Lagoons and reedbeds are used with Akalla and Emerson Flowforms (Figs. 7.21 and 7.22).

Kersty MacColl (Project 1106 U.K.)

Built in 1990 by Julian Jones of Elver Eco systems, in a private garden in Ealing, London, in which part of the Sevenfold II cascade is used. It is shown after planting began to develop. A very interesting small scale solution which enables water to be discharged into the ground to help replenish groundwater supplies to help prevent house subsidence (Fig. 7.23).

Highgrove Park (Project 1148 U.K.)

Built in 1991 by Uwe Burka in conjunction with a reedbed and willow filter bed for a decidedly fluctuating population. A specially designed cascade with vigorous movement was planned (Fig. 7.24) but in the end a single Scorlewald Flowform was installed to circulate the water within the post-treatment lagoon.

Body Shop factory, Sussex, U.K. (Project 1181 U.K.)

Built in 1993 by Jane and David Shields of Living Water, as a

Figure 7.21 (opposite above) and 7.22 (opposite below). Biological sewage plant for a small community on the west coast of Norway at Hogganvik which discharges into the sea. Three lagoons with filter beds and cascades are used. At the other side of the pool shown, water is discharged through a Phragmites filter bed to the final fish pond further down the slope.

Figure 7.23 (above). A small household sewage purification system in a built-up area of west London. Here shown when plants had begun to grow.

Figure 7.24 (below). A free kind of idealized sketch for a Flowform cascade intended for a Highgrove commission.

Figure 7.25 (above). For the Body Shop production facility at Littlehampton, Sussex, a research system has been used already over a period of some years, for dealing with waste products. This is under glass in a protected environment and provides an ongoing research object which is gradually being developed further. The Järna Flowforms were initially mounted too steep and were therefore not functioning optimally and also would do better with a higher flow-rate. With the right input and adjustments this could prove to be a very useful and important project.

research project using several Järna cascades to establish efficacy in dealing with toxic waste from production processes (Fig. 7.25).

Greece (Project 1493)

Built early in 1993 by Uwe Burka using small Akalla Flowforms for a fairly large community.

Trussocks Hotel (Project 1596 U.K.)

Built in 1993 by Iris Water for a hotel in Scotland using Malmö Flowforms.

Stensund (Project 1734 Sweden)

Carried out by Virbela Atelje and in operation since 1989, this is for an Ecological Research Centre within the Folkhögskola, Trosa, and is

Figure 7.26 (left). In the town of Kolding, Denmark a housing area is serviced with this fascinating sewage disposal plant. The pyramid greenhouse contains a commercial growing unit and is fed with water from the reed-bed system with a Flowform cascade and meandering stream connecting the various components.

FLOWFORMS: THE RHYTHMIC POWER OF WATER

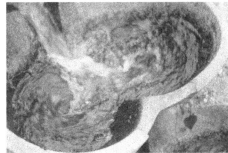

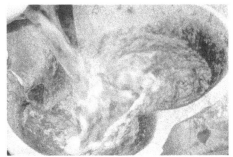

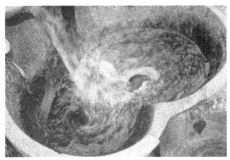

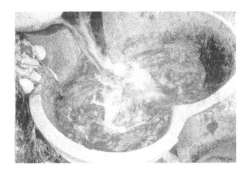

designed with public access in mind. Flowforms are used to treat the discharged effluent as it leaves the plant before entering the Baltic.

Kolding (Hanne Keis Denmark)

Malmö Flowforms incorporated in an extensive project built during 1994 to treat sewage from a housing complex in the town. A glasshouse pyramid contains commercial growing beds which use the treated water passed through filter beds and Flowforms (Fig. 7.26).

Herrmannsdorf (Project 1756 Germany)

The Vortex Flowforms were built into the system in Spring 1994 to function at the end of the first phase of treatment. The cascade discharges into the extensive post treatment ponds as shown in the pictures (Fig. 7.27). Detail phases in the movement within one of the Vortex Flowforms (Fig. 7.28).

Figure 7.27 (above). Herrmannsdorfer Landwerkstätten is an organic farm with a number of processing units including cheese and milk product curing cellars, bakery and brewery in each of which Flowforms are in use including this final water purification treatment plant.

Figure 7.28.(above). Details of the movement taken in one of the Vortex Flowforms over a short period of time. The cascade discharges into the extensive post treatment ponds as shown in Fig. 7.27.

Withington (Ebb & Flow U.K.)

Built during 1995 for a small hotel with lagoons and cascades using Malmö and Emerson Flowforms.

Slott Tullgarn (Project 2565 Sweden)

This is a large project for the fluctuating population in a royal summer residence and visitors' centre, built by Rune Östersen in 1996 with the collaboration of Nigel Wells, our associate who constructed the Flowform cascades. It consists of lagoons with the large Scorlewald Flowforms in two places, two Akalla cascades and a Järna cascade. Composting beds are also included (Fig. 7.29–7.31).

Water Garden Chengdu (Project 2672 China)

Planned and carried out in 1997 by Betsy Daemon (a US landscape architect) using water from the river, to improve its quality through biological treatment, and Flowform Cascades as a demonstration of river improvement. The Chinese talk of this installation as a model for China, which remains to be seen. Further ideas were later discussed in Shanghai.

Figure 7.29–7.31. Winter operation showing several cascading Flowform systems at Slott Tullgarn. In the background below, the sludge plant filter beds can be seen. These can be flooded intermittantly over a ten year period, the liquids are drained through and the solids dehydrated and detoxified to form humus which can eventually be removed for use.

8

The next generation of Flowforms

Järna Flowform

The Järna Flowform started out as the Basic Flowform (see Fig. 7.2), which was the first deep form with a flat base and more or less conical walls generating a relatively vigorous movement. It was intended primarily for functional purposes. The joints were semi-butting, that means a ball and socket joint, allowing some lateral movement, providing some overlap but not demanding a drop in the total height. This reduced the gradient demand, each step being about 6.5 cm. The design was amended at this stage with Nigel Wells. The Järna Flowform used initially in the context of biological sewage treatment (Figs. 7.12, 7.19, pp. 101, 103) has proved an excellent tool for mixing liquid preparations in organic biodynamic farming (Fig. 8.2).

Figure 8.1 (below left). The Järna Flowform started out as the Basic Flowform (010670). In 1973 the first cascade with basic Flowform I took from England to start off the process. All the Flowforms in this summer were made in cement Fondu (a quick setting alumina cement) using fibre-glass and sand for the shell casts made from rubber moulds.

Figure. 8.2 (below). A fine picture by Andrée Brett of the Järna Flowform showing an excellent function, with John Pearce in New Zealand. He was one of the pioneers in the use of Flowforms for mixing liquid Biodynamic Preparations in New Zealand. This is on the cover of the book Biodynamics N.Z. 1989.

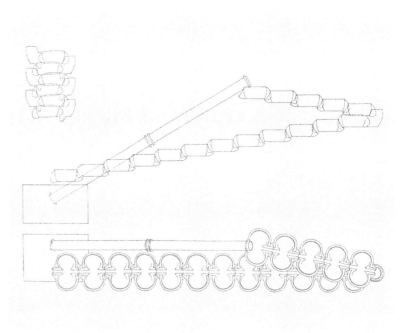

Figure 8.3 (above right). A Järna Cascade built in conjunction with an Archmedian Screw, to treat irrigation water or mix Preparations. The small additional unit facilitates vertical stacking or a 170° change in direction.

Figure 8.4 (above). The Järna Flowform used in a straight channel in Park Zijpendal, Arnhem, to enhance water movement which would otherwise be totally laminar.

Some time later an extra unit was designed to fit on to the exit which enabled the Flowform to be stacked vertically and for a cascade to change direction by 180° or by 170° when used for instance with an Archimedean screw (Fig. 8.3).

I was able to use this unit to good effect for irrigation purposes in a biodynamic garden in Tenerife. The land is terraced and channels are arranged down the slopes. The terrace walls could be negotiated by using the stackable units to bring the cascade down to the next level.

The last major amendment was made by Nick Weidmann in the early 1990s. This necessitated a change of the base so that it is now mounted simply on a slope of 18% and this simplifies installation significantly. Also the design of an entry unit combined with the Flowform obviates the need for an expensive special entry.

Another useful application consists in building Flowforms in a straight channel through which water would otherwise tend to surge in a typically unnatural way, to bring it into a rich rhythmical movement. Here an example is shown in Park Zijpendal Arnhem, using the Järna Flowform (Fig. 8.4).

Emerson Flowform

This started as the Open Flowform. It was a period during which many design experiments were undertaken, and only later was the first fibreglass mould made, in May 1974. The form might well have been inspired by the Torso Flowform which was the first experiment made with path-curve surfaces (see Fig. 5.17, p. 74). The idea was to build a rather flat vessel over the front edge of which water would spill alternately in a waterfall. Sometime during May 1974 three of the forms were built as a cascade in the garden pond at Emerson College (Fig. 8.5). An exciting phenomenon presented itself the next winter when the surface ice was seen to thaw in a way which revealed water movements in the pond beyond the cascade (Fig. 8.6). Also during this period an education exhibition was planned in Stockholm at Liljevalchs. For such purposes this kind of Flowform was needed but one that would not spill over the forward edges. Thus a further development brought about the form which also contained the water efficiently

Figure 8.5. The original Emerson Flowform Cascade, set up in a garden pond on the campus. This was still the early version over the front edge of which the water was to cascade alternately.

Figure 8.6 (below top). This is one of the surprising pictures resulting from Flowform movements. Otherwise completely invisible movements below a cascade, were shown up by virtue of the melting ice, observed over the first winter as the Flowform was left running.

Figure 8.7 (below). In order to prevent spillage over the front edge, for interior use for instance, I built a rim which contained the water during function.

Figure 8.8 (right). The Emerson Flowform has been used very widely for landscaping projects, here in a local garden in Sussex at Cowden.

while maintaining the same quality of movement. The amended cascade was mounted for a while in the water gardens at Järna (Fig. 8.7) before being eventually used for other purposes on a farm. The Emerson Flowform is used widely for landscaping projects, for example at Cowden, Sussex (Fig. 8.8).

Acrylic Flowform

In order to experiment with transparent Flowforms through which one can observe the shaded movements of the water from below (Fig. 8.9) it was decided to work with acrylic material. A plaster prototype had to be prepared from which to make a very heavy duty epoxy negative to be fitted into an air pressure press. This was carried out by Rudolf Dörfler in Dornach. A quite thick perspex sheet is preheated, placed in the press and blown over the mould.

When working with plastic in this way, a very smooth surface is inevitable. Water slips over it rather easily and this makes the process exceptionally sensitive.

Such Flowforms are visually exciting and provide a great attraction in the exhibition context especially if used overhead so

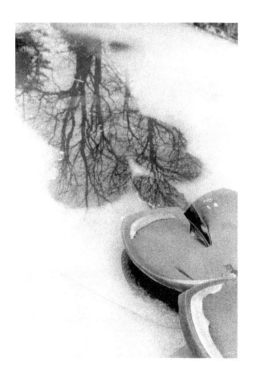

that water movements can be experienced from below. In order to document this phenomenon white translucent material was used for photographing and filming (Fig. 8.10).

A transparent smoked material was used for a very effective project in an open plan office in Berlin (Fig. 8.11).

Malmö Flowform

It now became necessary to work on a more robust larger flow-rate Flowform (250475) which was destined for much use in biological purification systems. The first cascades however were used extensively for exhibition purposes, in a Waldorf Education context in Malmö; at the Stockholm Modern Art Museum; in Dornach, Switzerland, and at Emerson College (Fig. 8.12). The

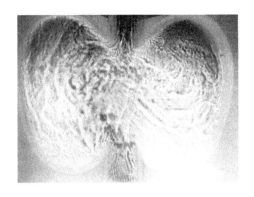

Figure 8.9 (above) (290874). The so-called Acrylic Flowform made in transparent or translucent material. This enables observation from below of the wonderful shadow play of the water as it swings through the lemniscatory vortices.

Figure 8.11 (left). The set-up in our farm studio for photos and filming. I am looking forward to demonstrating this more often when we have improved facilities for our work.

Figure 8.10 (below). This was a very fine project for open plan offices in Berlin. Unfortunately it was later dismantled and replaced by another material, possibly due to the deposition of calcium. But this problem could certainly have been avoided with correct maintenance.

Figure 8.12 (top left). The Malmö Flowform, first used for an education exhibition in south Sweden is seen here at Emerson College on May Day '76.

Figure 8.13 (above). Mounted in Järna after the Ararat Exhibition it is still functioning after twenty-seven years all year round. Often in winter all manner of ice formations can be experienced.

Figures 8.14 and 8.15 (left). In summer two pictures give a good impression of the vigorous movements generated. This Malmö Flowform cascade is in the fourth biological pond in Järna.

first permanent cascade was built in Järna, this has run continuously, winter (Fig. 8.13) and summer (Fig. 8.14), since 1976. Two photographs are used to indicate more clearly the vigorous movements which are generated. The Director of Skansen Zoo in Stockholm told us the algae growing in these Flowforms proved that the quality of water in this last pond was very good.

It was built into a fish farm in Sussex. Here, I think for the first time it was possible to observe the fish especially congregating below the exit within the rhythmical stream, in preference to a waterfall issuing from a pipe nearby. In this instance fish were lined up three or four metres. This occurrence was observed in a number of widely dispersed cascades and led to the idea of implementing cascades in the fish-ladder context.

As the result of a practical conference at Sundet, Mösvatn, in Norway, Flowforms were cast and mounted in a nearby stream with the view to encouraging the fish up from the lake (Fig. 8.16).

The Malmö Flowform took on a significant role as part of the

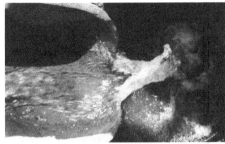

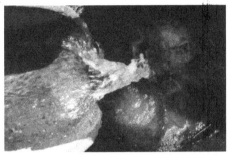

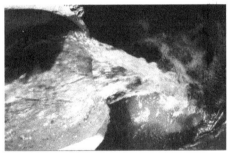

complex research installation at Warmonderhof (see Fig 10.5 on p. 137) where as the middle cascade it was used for the comparative tests with the Step cascade. One especially interesting feature is the exit of the terminal form as a special lip which encourages the discharge oscillation to form a waterfall landing in a lemniscatory movement. Fig. 8.17 gives some impression of the movements.

The design reached its present and final form in 1985 through the work of my colleague Nigel Wells.

Akalla Flowform

In August 1975, the landscape architect Lars Crammer visited Järna, and immediately I heard from Arne Klingborg that a project for a children's recreation area north of Stockholm was likely. I visited the site in Akalla on October 31 and discussed details with Lars Crammer and others over the next days. Confirmation of the project was given on November 5, 1975, the day of my return to England.

As a tribute to our clients and as an indication of what is involved in such a project I would like to describe in some detail how this project ran. It proved to be an example of a rather ideal process. It was perfectly timed and came as an immediate result of my efforts in Järna. One could not have wished for anything better at the time, as support for this new impulse.

In Akalla, the sloping rock-strewn site suggested Flowforms of a

Figure 8.16. As the result of a conference in 1978 at Sundet on Mösvatn in Norway with Unni Coward, we cast and built a cascade in a mountain stream. Our children take great pleasure in having a good drink!

Figure 8.17a–c. A special end Flowform used to terminate cascades (see also Figure 5.14 on p. 71). This lip employs the oscillating process to emphasize the movement of the waterfall which lands in a figure-of-eight.

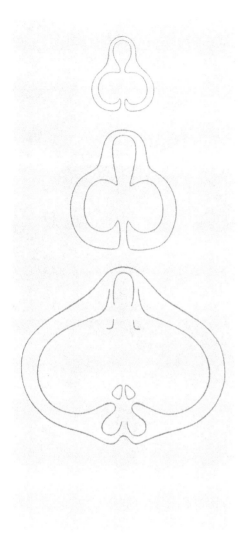

boulder-like shape. My idea was to have these in three sizes to accommodate the irregularity of the slope and to present vigorous and calmer movement in the smaller and larger forms respectively (Fig. 8.18). We wished to bring as little disturbance to the area as possible. Therefore the border between rocks and grass was chosen.

On November 12, I started the first scale model and by the end of December with Nigel Wells we had the large Flowform of 180 cm diameter, working. On December 19 and January 10 we began the middle size (90 cm diameter) (Fig. 8.19) and the small size (50 cm diameter) respectively.

Plaster casting and final working up continued, followed by moulding for concrete casting; at this point we were joined by Ian Corrin, a craftsman from Wales. Technical details for the installation were worked out in general and communicated to Sweden there to be finalized for the site work.

We made extra polyester casts of each Flowform as lightweight maquettes to enable us on site to plan the positioning of the Flowforms in very difficult terrain.

The three of us left for Sweden at the beginning of March. We arrived in Märsta by car with all our moulds, on the 8th. Märsta was the location of a prefabricated concrete-unit casting factory owned by the builders Ohlsson and Skarne AB, and here all facilities were made available to us for casting the Flowforms.

Figure 8.18. The Akalla Flowform shown to scale in its three sizes. Naturally the same volume of water is used so this factor demands small Flowforms with deeper profile demanding steeper gradient, to larger Flowforms which become gradually flatter and demand less gradient. This was one of the reasons for creating the metamorphic sequence; to negotiate changing gradients.

Figure 8.19. The clay prototype of the medium sized Akalla.

120

Figure 8.20. *Removing the mould from the cast of a large Akalla.*

Figure 8.21. *Turning the cast over using a special extension built onto the mould so no damage is done to the fabric of the mould.*

Figure 8.22. *Preparing the mould again with steel armature for the next pouring.*

Figure 8.23. *Ian Corrin waiting for the mix to arrive by transporter crane.*

Figure 8.24. *Concrete about to be poured into the mould and vibrated.*

Figure 8.25 (above). My two colleagues: Nigel Wells (left) and Ian Corrin (right) who passed away in 1993 after a prolonged illness.

Figure 8.26 (right). With repeated testing with water to ensure the levels were correct for the function. Kersti Biuw joined us from Stockholm to help.

Figure 8.27 (below). The workshop in France where in 1987 moulds for the Sevenfold II were made and new moulds for the Akalla destined for Vidaraasen in Norway.

Each morning early we demoulded, the back of the mould being removed (Fig. 8.20) to prepare for the difficult task of rotating the heavy cast with the help of a crane (Fig. 8.21) and moving it to storage in the yard. The next task was to clean out the moulds in order to cast the next set from all three sizes.

When all casts were ready at the factory, they were transported to the site where we had continued setting up positions and levels.

One by one the Flowforms were placed, (Fig. 8.26) with repeated testing with water to ensure the levels were correct for the function. This is always very critical and one cannot rely on measurements alone. It was certainly exciting to experience the wonderful dramatic ponderous movements of the large Flowforms compared with the vigorous rhythms of the smaller Flowforms.

By April 1 we had finished and commissioned the cascade.

In 1987 new moulds for the Akalla series were made in France by Jan Gregoire with Nick Weidmann helping. These were destined for the new Flowform casting workshop initiated by Lars Henrik Nessheim and his colleagues at Vidaraasen Landsby, a village community for the handicapped in Norway. Many Flowform designs have been cast here (see Fig 12.29 on p. 162).

9

The Metamorphic Sequence

In Chapter 6, I discussed how my central theme became to develop a metamorphic sequence which offered a spectrum of rhythms and gestures to water flowing through. The sevenfold sequence was suggested by the way nature builds organs for fluid processes, as well as by the natural harmonics of wave sequences. The question was: could we offer in *space* the 'unified dynamic gesture' that characterizes the metamorphic development of organs in *time*? Table 4 outlines some of the working principles that underlay, and continue to inform, this development.

Table 4: Life-processes related to qualities that might be developed in a sevenfold cascade.

Life processes	Equivalents relating to a cascade
taking-in (breathing)	*entering*
acceptance (warming)	*relating to surfaces*
digestion (nutrition)	*rhythms*
secretion	*receptivity of water's innermost structure*
nourishment	*oxygenation and planetary influences*
growth	*subsequent support of growth processes*
reproduction	*subsequent organism regeneration and enhancement*

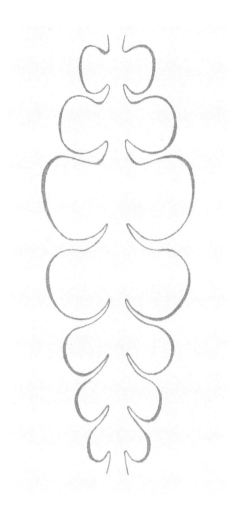

Figure 9.1. The form of the first experimental sevenfold cascade.

If the same Flowform design is repeated in series, like-rhythms can compound to form an increasingly complex rhythmical pattern as water proceeds through the cascade. However this pattern naturally maintains the signature of the single Flowform. Individual Flowforms in such a cascade can be observed to fill up and empty out over a longer period. That means a vigorous period of movement develops when the form is 'full' or functioning optimally, followed

123

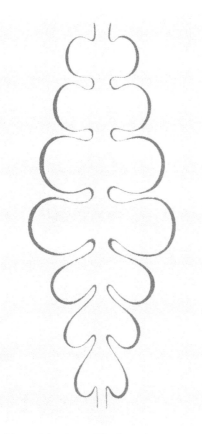

by a calmer period when it is practically empty, while the individual frequency tends to remain relatively constant.

Flowforms of varying size and identical flow-rate generate a larger range of rhythms within the cascade. Smaller Flowforms generate vigorous, fast, more spherically contained and three-dimensional movements, while larger forms can generate slower, calmer, more two-dimensional movements. The cumulative effect can be very complex but nevertheless consistent. The function of such a complex unit is to offer water a manifold and rich spectrum of treatment through an ordered metamorphic quality of exposure approaching to some degree the fertile complexity of nature's own living realm.

The first sevenfold cascades

Early design work

Over the first fifteen years many attempts were made to carry out the project. The first sevenfold experimental cascade (Sevenfold 0) was made shortly after the work with Flowforms began in 1970. The diagram (Fig. 9.1) shows the actual plan of the model, about two metres long, which was constructed with lead-strip walling on a flat metal sheet (Fig. 9.3). Thus the whole channel lay on a continuous slope. It was of course here not yet fully worked through but there is already a fairly clear indication of the angular sweep of the lemniscate in the individual Flowforms from the beginning to the end. Here, the greatest expansion is not yet in its appropriate position at the half-way stage. Later the original sequence was amended to bring the largest into the fourth mid-position (Fig. 9.2). Having an

Figure 9.2 (above). The first adjustments were made here, now the maximum expansion occurs in the fourth vessel.

Figure 9.3 (below). An investigative model of this sevenfold sequence, made with lead strips on a sheet of metal. A wall surrounds the whole to avoid leakage (see Figure 9.1).

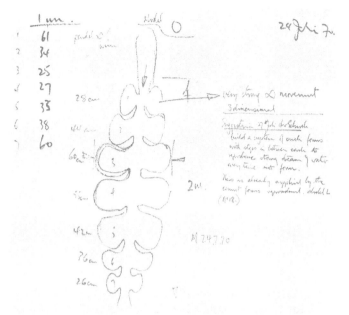

actual model made it possible to get an idea of the range of rhythm frequencies generated by the spectrum of sizes. These were noted (Fig. 9.4). These values relate to the overall dimensions of the Flowforms and it is these relationships which still require investigation in order that some specific harmonic set of associations be found for such a sequence.

Olympia Flowform

This Flowform was originally designed and submitted for an urban project in Nyköping, Sweden, where from a total of nine submissions it was accepted in principle by the Jury but in the end not commissioned as they were unfortunately unable to decide how they really wanted to carry it out! The project was later completed and first used at the Olympia exhibition centre in Earls Court, London, from which it takes its name (Fig. 9.5).

It was conceived not as a simple linear sequence but as a wider flow pattern displaying overall symmetry, incorporating seven different shapes of Flowform — three symmetrical and four asymmetrical (Fig. 9.6). After visiting the intended Nyköping site, I decided the area needed a compact cascade and devised a three-armed layout with a central entry. The main axis would be

Figure 9.4 (above left). Notes on the frequencies measured on the model.

Figure 9.5 (above right). The Olympia Cascade first exhibited at Earl's Court in London. 1977

Figure 9.6 (below). Original line composition of what became the Olympia complex of seven Flowforms, three symmetrical and four asymmetrical.

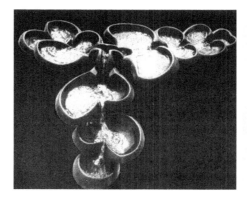

Figure 9.8. The executed design of the Olympia complex.
Figure 9.9. Plan drawing of the Olympia complex showing the three end vessels which feed back into the main reservoir beneath, from which water is circulated.

Figure. 9.7. Principle showing the change of gesture in the lemniscate as the lobes move their emphasis in the direction of flow. The range of angular movement or sweep indicated in the vessels, inspired by the various phenomena described above, influences the lemniscatory process as illustrated above, and described as follows.

In the first form the oscillating stream is carried dramatically backwards alternately to the left and right, only to swing forwards again into the main stream which it crosses at the front of the vessel. In the middle form the lemniscate swings out to the left and right perpendicular to the main stream, the crossing point is moving from the forward position towards the rear. In the last form the oscillating stream continues forward to the left and right, only to return towards the centre, crossing the main stream at the rear of the vessel.

In the sevenfold cascade complex, the lemniscate makes three steps towards the middle condition and three steps beyond it, performing a swinging gesture the changing angle of which accompanies the direction of flow.

dominant, with symmetrical Flowforms, while the two side arms would be shorter and be made up of asymmetrical Flowforms large and small, left and right handed. In order to clarify the composition as a first stage I made a flow diagram. This drawing indicates three symmetrical Flowforms, large medium and small. The character of the lemniscate changes in that the lobes change emphasis from the rear to the fore (Fig. 9.7). In the asymmetrical Flowforms the lemniscates are naturally also asymmetrical.

After the sculptural design was carried out (Fig. 9.8) the plan of the complex could be drawn (Fig. 9.9). With a view to simplifying the reproduction process the Flowforms were designed to splay out to an edge so that one-piece moulds could be used initially.

The two smaller asymmetrical Flowforms positioned at the ends of the two arms, discharge into the end vessels over a convex lip which creates a pulsing globe-like waterfall. The two main phases of movement swing in vessel number seven (see Fig. 9.6) are illustrated here (Fig. 9.10) in the small cavity fast, in the large slower.

Following the setback in Nyköping, we received another commission, this time for the first Mind Body and Soul exhibition to be held at the Olympia exhibition centre, Earls Court, London (see Fig. 9.5). Using the designs we had already elaborated, the new work was financed with the help of Martina and Christopher Mann and the full-size installation was completed during the winter of 1976-77 as a main entrance feature at the centre.

Eventually the Olympia model was used very effectively at the ING Bank in Amsterdam for a permanent garden feature. In plan, the complex of Flowforms unexpectedly provided a sculptural metamorphosis of the architectural plan of the Bank towers and it was for this reason that I recommended its use as a whole (Fig. 9.11). In a further garden area, three linear cascades were built, two of these discharged from a good height into a large pond as pulsing waterfalls (Fig. 9.12). In this way a unifying motif was maintained through the gardens by using the same Flowform design in quite different orientations.

The four asymmetrical Flowforms indeed became the ones most used and for some time were in production for a number of projects in various combinations: for instance in Driebergen, Warmonderhof (see Figs. 10.2 and 10.3 on p. 135), Haus de Vaal and

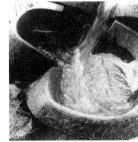

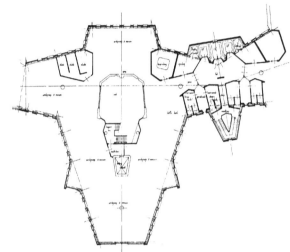

Figures 9.10 a&b (above right).
The two main phases in the movement generated in Nr 7, fast in the right side slower in the left side, at the exit a pulsing dome-like waterfall.

Figures 9.11a (right) and 9.11b (right below).
The whole Olympia used in one of the NMB (now called ING Bank) gardens in Amsterdam. It takes on the nature of a sculptural metamorphosis connected with the architectural plan motif of the ten Bank towers. Fig 9.11a shows the tower plan with the adjoining transitional section to the next tower with lifts and toilets. Fig 9.11b is a picture of the gardens with the cascade (compare Figures 9.8 and 9.9).

Figure 9.12 (below).
Another cascade using the asymmetrical Olympia Flowforms in three separate streams two of which provide pulsing waterfalls.

Figure 9.13. In a park designed for the blind in Ulm, Kolbengraben, Atelier Dreiseitl set up a three-fold cascade with Olympia Flowforms. The blind were interested especially in a water feature to which they could have access. They wished not only for an acoustic effect but one one which they could experience with their hands.

Figure 9.14. Olympia Flowforms used for a private garden in Switzerland set up by Atelier Dreiseitl. Earlier in the UK we had designed an intermediate vessel so that the small asymmetrical Flowforms with the pulsing dome waterfall could be used as a cascade.

Klazineveen, (Holland); Würenlos (Switzerland) (Fig. 9.13); Norrköping Sjukhuset, (Sweden); Engelberg School, Ulm Kolbengraben (Fig. 9.14) and the Kunstgewerbemuseum, Berlin (Germany).

Research models

Already in the mid-seventies scale models were made for a sevenfold Flowform sequence of about four metres long.[13]

Progress was also made with the idea of incorporating the 'rhythm enhancement effect' for which a large detail model was constructed. This effect has to do with the influence of rhythms, to be carried over from one Flowform to the next, which could enhance the dynamic of subsequent rhythms over periods of longer duration. This is shown here schematically in Flowforms of different size. The smaller demonstrates a quicker rhythm, the larger a slower rhythm. The relationships naturally shift in time. Discharge from the first Flowform oscillates from side to side, periodically intercepting and briefly cancelling the movement in the second Flowform only in a later phase to enhance the slower rhythm substantially (Fig. 9.15). The

degree of influence carried over from one Flowform to the next is regulated by the form and composition of the interconnecting channel, this can be very critical to the functioning of subsequent Flowforms even after a number of steps.

These extended rhythmic periods which can lead to recurring high points remind us of the observation that after a given number of waves reaching the coast, each seventh can be the largest.

Further parameters, to name only a few of the possible directions of research, had to do with the harmonics of the individual Flowform dimensions related with each other, the detailed quality of the surfaces involved and the progressively changing design gestures.

In relation to design, as well, there was the constant question regarding methods of circulation as we wanted to find other solutions rather than the use of a pump and considered ways in which, for instance, we could wrap the cascade round an Archimedean screw!

Design of the Sevenfold I

It became possible to carry out the first complete sevenfold project with Nigel Wells in 1985, albeit on a smaller scale than anticipated and in very simplified terms due to limited resources.

Sevenfold I consists of three cast units, the first includes the entry with numbers one and two; the second, with numbers three, four and five; the third with numbers six and seven. The whole is fixed in one slight 'S' curve plan (Fig. 9.16). Later on after this com-

Figure 9.15. The diagram again which shows the principle of the rhythm enhancement effect.

Figure 9.16. The Sevenfold I Cascade with flow diagram indicating the changing gesture of the lemniscate.

Figure 9.17. The Sevenfold II Cascade which was developed stage by stage from the first design. The Flowforms are now individual with ball and socket joints which enable the plan to be curved both left and right. The gradients also change from steeper beginning and end to a more horizontal middle section.

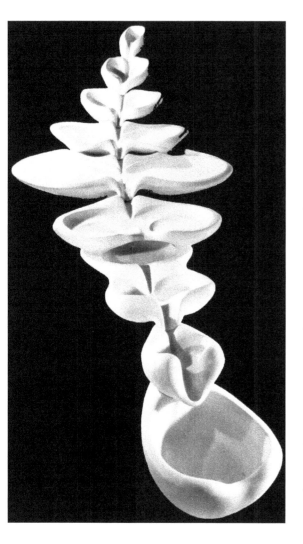

Figure 9.18. The original plaster prototype designed in conjunction with Hansjörg Palm and in the final stages with Nick Weidmann.

plete unit went to the U.S.A we began working on the separation of all the forms, with individual joints to allow for maximum flexibility. This led in turn to the next project for which finances began to appear.

Design of the Sevenfold II

During 1986 with Hansjorg Palm further design features were evolved. The Sevenfold II series of separately jointed Flowforms was developed step-by-step out of the first with the flow-rate remaining constant throughout (Fig. 9.17). The very nature of a smaller Flowform with its vigorous movement demands a steeper gradient and more spherical or three-dimensional design. So here we show a drawing by David Joiner of the elevation with changing gradient (Fig. 9.19).

As the Flowforms grow in size, so less gradient is needed and the movements are slower and more ponderous. We decided to try a more enclosed form with re-entrant surfaces for numbers one and seven. Although this complicated the business of reproduction it had to be tried at some juncture. This deep inner cavity enables the movement to be very vigorous but contained. The lemniscatory movement is thus more three-dimensional.

Almost immediately through Anna Pauli a potential client appeared, who was to finance the completion of the mould-making process for installation in an exhibition in Switzerland. Although the Swiss project had to be called off half way through, we were able to complete the moulds as this work was well on the way. The first casts were in fact exhibited in Stuttgart.[14]

Summary

This theme of an 'organ of metamorphosis' for water is central for the whole concept of the Flowform impulse. It has met with very much interest and up to the present has been used for at least one hundred projects in fifteen countries round the world (see Chapter 12).

Cascades exhibiting metamorphic tendencies

Sevenfold 0

This relatively simple model was made with lead walling, for easy manipulation, on a flat metal sheet; it was an initial attempt to try out the idea of an 'organ of metamorphosis.' Its general plan of an expand-

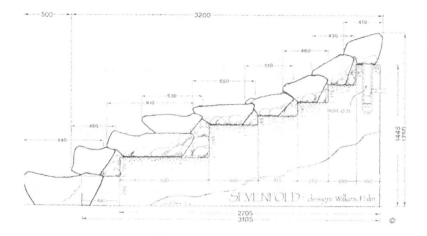

Figure 9.19. An elevation and foundation diagram of the Sevenfold II Cascade. Illustration: David Joiner.

ing and contracting series was based on the metamorphic transformation of a single Flowform with its narrow entry expanding into the left- and right-handed cavities and final contraction through the narrow exit (Fig. 9.2). As described in Chapter 9, the lemniscatory process also goes through a gradual transformation (see Fig. 9.7).

Random cascade

For this first project in the Järna lagoon system (see Chapter 7), a cascade was built on a long beam using Flowforms of three different sizes (Fig. 9.18). These are variously named as Järna, Slurry and Sewage Flowforms. They are used in random order.

Akalla

A landscape architect from Stockholm was inspired by the Järna installation to commission a cascade for the children's recreation area at Akalla, at that time a new high-rise housing project. Akalla Flowforms were designed to fit into a boulder landscape (see Chapter 8 for a detailed account of this project).

Olympia

The whole Olympia composition was conceived as a metamorphic whole of relating elements rather than a linear sequence of individual Flowforms. The original concept (290776) was recorded in the form of a drawing (Fig. 9.6) while on holiday in the Norwegian

Figure 9.20. It is possible to use the Sevenfold II Cascades in certain random variants, here is shown 0,5,6,5,6.

mountains! It is a fascinating activity to imagine and conjure up a picture and whole orchestration of ordered water movements.

Sevenfold I

Having had the original idea during the first days after the discovery of the method, it was fifteen years before we were able in 1985 to design the first sevenfold sequence as conceived. In the meantime many ideas had been worked through and these formed some basis for the actual execution.

Sevenfold II

The second design sequence followed immediately in 1986 and was developed step-by-step out of the first. As one Flowform has a narrow entry and exit so the whole series begins and ends with small Flowforms, expanding in size towards the middle. The smaller forms demand increased gradient, so the beginning and end is steep, the middle more horizontal. As well as the change in the form of the lemniscate the sculptural gestures also move through a spectrum of change.

Some time later when we were making a number of polyester positives as 'mother-moulds' we were able to try out different combinations of the seven Flowforms (Fig. 9.20).

Sevenfold III

Flowforms 1 and 7 prior to this have re-entrant surfaces; it was necessary to try this out at some point but not so easy to mould and cast, so it was decided to change these two designs. The entry and 1 are now combined to 0/1 so this provides entry but also functions as a Flowform. Number 7 is also opened out to facilitate casting.

10

Research with Cascades

In the early 1970s, scientific circles seemed to show very little understanding, or even interest, regarding the influence of rhythms. In areas surrounding research on Flowforms, however, a perception of rhythms has been a constant part of our everyday awareness. Having made the unexpected discovery, through working with the proportions of vessels, how rhythm is generated by means of resistance, the next set of questions concerned what effects rhythms could have on water, and especially upon its life-supporting capacity. Thus the quest began, to try and find out what might be happening to water — in a qualitative sense — under this kind of influence.

While investigators like Theodor Schwenk and later Nick Thomas (unfortunately George Adams was no longer alive when this particular work began) were enthusiastic and serious about exploring the effect of rhythms on water, other experienced scientific thinkers would say: 'Yes, the Flowform is a beautiful and exciting phenomenon from an artistic point of view, which you must continue to work with and develop, but forget the whole question of qualitative effects!' This has proved an incorrect assessment and as time goes on many scientific and other professionals have begun to take the phenomenon of rhythm and its effects much more seriously. Indeed it must be increasingly evident that rhythms play a major role in all human activities.

To name only a couple of critical areas, we have homeopathic and anthroposophical medical practitioners, together with companies such as Wala and Weleda which prepare medicines, as well as biodynamic organic farmers, who see rhythmical processes as fundamentally necessary to life. In the realm of education, rhythm is of the utmost importance in learning experiences and this is practised in Waldorf schools particularly. In exploratory areas of science, related interest is now being shown in dynamic systems and chaos theory, as well as principles of self-organization and complex generative patterns. However, the scientific world at large is still extremely sceptical and often aggressively opposed to such views because the power and truth of the arguments are literally feared. [15]

Scientific background and investigations

Though always limited by financial resources, the research group that gradually formed over the years has tried to maintain a range of scientific activity and investigation, particularly in the area concerned with rhythm and metamorphosis. Here, work on Flowforms has been inspired by a specific scientific attitude to natural phenomena, as exemplified by Goethe's approach to observation and perception.[16] One of our main concerns was whether the life-sustaining properties of water can be enhanced or regenerated through the rhythmical lemniscatory movement induced by Flowforms.

Besides their aesthetic qualities, Flowforms do appear to have significant ecological and environmental applications. A comparative study undertaken by C. Schönberger and C. Liess in Überlingen, Germany, on research articles about Flowforms indicates that the properties of water are altered if water is run through a cascade of Flowforms. Penetrated by rhythmical movements, Flowform-treated water not only becomes highly oxygenated but also supports rhythmic biological regenerative processes more intensively.

The influence of parameters such as temperature, density, flow rate and viscosity on the rhythmical movements of water were studied in a cascade of four Flowforms at Luleå Hoegskolan in Sweden (Strid 1984). Water density was altered between 1000 and 1170 kg/m3 by addition of salt, and viscosity was altered with a powder of polyeteneoxide in the range of 10^{-6} to 10^{-2} m^2/s, and temperature ranged from 5 to 48°C. The frequency of the water pulse in a Flowform was found to depend only on the quantity of water, but not on temperature, density and viscosity of the fluid. In one particular Flowform, for example, the pulse started at 3.0–3.6 lit./min with a frequency of 104.0 min^{-1} (1.73 Hz) whereas at 7 lit./min the frequency was 107.6 min^{-1} (1.79 Hz).

In order to measure electromagnetic properties, electrodes were immersed in the water at different positions within the Flowforms. It was found that the voltage in the Flowforms pulsated with the same rhythm as the water (Strid 1984). This enables rhythm frequencies to be measured accurately (see Appendix 3).

The properties of Flowform-treated water generally suggest the value of this approach for waste-water treatment. In a pond system for waste-water treatment in Solborg, Norway, the oxygen content of Flowform-treated water increased from 30% to 90% (Trond Mæhlum 1991). The continuous rhythmical movements induced by the cascade also prevented freezing of the pond in winter.

The transmission coefficients for oxygen in different types of Flowforms and in a step cascade are very similar: Olympia 0.49; Malmö 0.45; Järna 0.39; step cascade 0.46 (de Jonge 1982).

The main issue nevertheless remains the influence of a strongly induced rhythmical environment upon biological processes, and here the comparative research undertaken with the installation at Warmonderhof in Holland has been of great importance.

The Warmonderhof Project

Much interest was generated by the Järna water treatment installation (see Chapter 7) and as a result Professor Jan Diek van Mansvelt of Warmonderhof Farm School — near Tiel in Holland — came to England in June 1976 and met with Professor Herbert Koepf, Nigel Wells and myself at Emerson College. His enquiries led to our building an installation for waste-water treatment at Warmonderhof. Effluent from the school, which farms on biodynamic principles, was normally discharged via a septic tank into the surface water drainage canal surrounding the complex. This in turn discharged into the nearby river, which was far too overloaded, static and thus eutrophic for the arrangement to be acceptable in such close proximity to the buildings. The purpose of the new installation (Fig. 10.1) was to create a kind of digestion organ for practical as well as educational purposes.

The design was intended to provide facilities for comparative research work, using various treatments in two separate sequences of lagoons which were to be as similar as possible. The channels were to be constructed quite shallow and the lagoons rather deep. This would provide varying flow conditions and places where the water would remain over longer periods. Thus plants demanding different habitats could be accommodated. The ambitious plan was developed to mount three Flowform cascades and one simple step cascade. All foreseeable combinations were built into the system.

Looking from above (Fig. 10.4) with the step cascade visible on the lower left of the picture, one can see clearly the inner series of lagoons leading to the rectangular basin, right foreground, which was planned

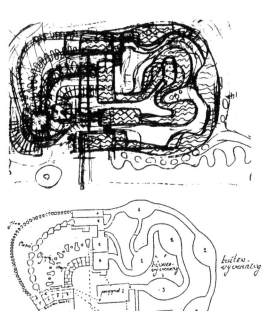

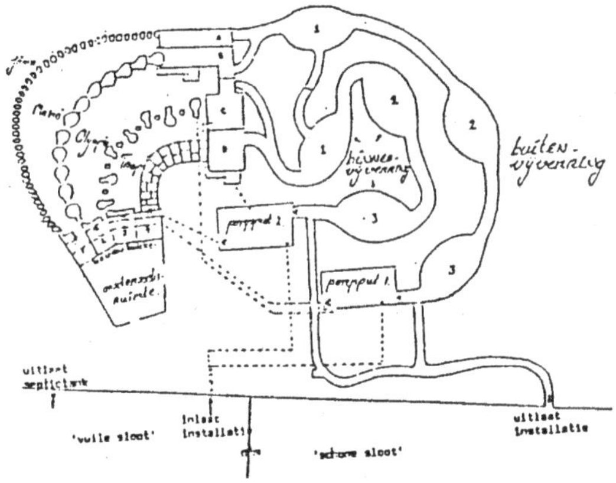

Figure 10.1. Opening ceremony of the Warmonderhof installation on 27 Sept 1980 with Louis van Gasteren officiating and Prof. Jan Diek van Mansfelt by his side. Louis van Gasteren co-founded Artec, a funding organization. Funds were acquired from different sources over an extended period: a contribution from Artec; a Government grant; and National Lottery funds.
Figure 10.2. My original sketch plan for the research complex.
Figure 10.3. A drawing made of the installation as it was carried out by Stef Hekmann.

Figure 10.4. View from above showing clearly the ponds and channels of the two separate systems with which comparative research was carried out.

to house an Archimedean screw; likewise to the far right a rectangular basin intended for the second screw. In order to move ahead without further delay, however, conventional pumps were used.

The research work was to examine a number of plant species in terms of the influence upon them of the laminar/gravity dominated movement in the step cascade (Fig. 10.5), compared to the rhythmical lemniscatory movements of the Flowform. For this purpose the centrally mounted Malmö Flowform cascade was used (right centre). In order to ensure the light conditions for all cascades were as similar as possible, the relatively high walls along the step cascade were removed.[17]

Another aspect of the research in Warmonderhof was to compare the more subtle variations likely to be brought about by the different Flowform designs. Which Flowforms might be more effective and for

which tasks? Particular rhythms and varying qualities of movement might be more or less suitable to support different organisms.

The basic research continued over a number of seasons between 1981 and 1984, during which a number of detailed reports were published. Regular observations were made in the two parallel series of three ponds, one series fed by water flowing through the step cascade, the other fed by water flowing through the Malmö Flowform cascade. Specific locations in both series of ponds were furnished each spring with selected water/bank plants from natural sites at an early stage of their annual development. It was noted in both series that macrofauna developed spontaneously out of the treated water, the root systems of the imported plants and from natural influences such as wind, birds, and so on.

Figure 10.5. A view of the installation with the planting gradually becoming established. It was initially intended not only to compare the Step Cascade with the Flowform Cascade but also to compare the possible different effects of the various Flowform designs with each other.

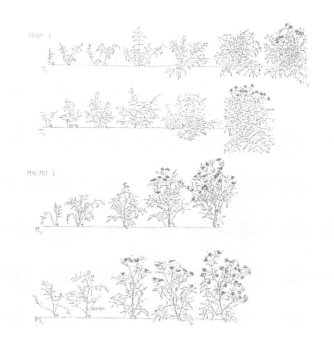

Figure 10.6. One set of results showed for instance the difference in growth of the same plant species in the two systems. The upper two rows show the strong vegetative result of the Step Cascade while the two lower rows show the generative tendencies resulting from the rhythmical Flowform treatment i.e. increase in blossom and seed production.

Professor van Mansvelt's final comments on the investigations carried out over the four years indicated differences between the two methods of treatment (van Mansfeld 1986). Generally speaking, plants grown in the lagoons treated via the step cascade tended to show increased vegetative growth, while plants grown in the Flowform treated lagoons exhibited an increase in blossom and seed production.

Step cascade system

Plant development was biased towards vegetative growth, (Fig. 10.6, top two rows) stressing wide leaf production, somewhat in the way of plants growing in shaded sites or by an overshadowed section of the wide slow moving downstream eutrophic part of a river system.

Macrofauna composition (Fig. 10.7) was biased towards species that prefer a darker habitat (deep water and bottom dwellers) displaying a somewhat softer rounded exterior shape, slower movements and a life cycle that includes a flying stage, such as midge larvae.

Observations of the water showed a tendency to be cloudy with a musty and ammonia-like smell.

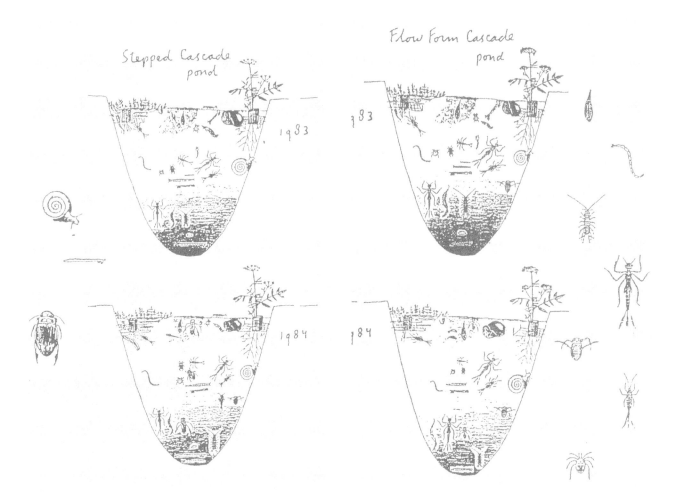

Stepped Cascade pond

1983

1984

Flow Form Cascade pond

1983

1984

Flowform cascade system

Here plant development was biased towards generative growth, (Fig. 10.6, bottom two rows) flowering earlier, colouring more deeply in autumn, stressing upright stems and smaller leaf production, somewhat in the way of plants growing in lighter sites or, for instance, by the upstream oligotrophic part of the river system.

Macrofauna composition (Fig. 10.8) was biased towards species that prefer a lighter habitat (upper water layers and surface), display a more pronounced and rugged exterior, faster and more nervous movements and a life cycle that remains in the water such as crustaceans and water mites.

Observations of the water itself showed a tendency to lower

Figure 10.7. With macro-fauna in the Step cascade system, the more mud-dwelling species proliferated in what appeared to be a darker environment.

Figure 10.8. With the Flowform Cascade a more lively situation developed in what appeared to be a lighter environment.

turbidity (clearer water) and a smell more like humus, eventually hay.

Incidental observations on goldfish behaviour showed them to be more playfully active and to prefer the rhythmically pulsing outlet of the Flowform cascade. They were slower and more passive when present at the evenly flowing — though similarly oxygenated — waterfall issuing from the Step Cascade. Similar preference for the pulsing outlet from Flowform cascades has been observed in fish at a number of other sites (a fish farm in Sussex, U.K.; Canstatt park installation in Stuttgart; and Büssnau Sewage plant).

Overall assessment

In this project, as we have seen, the life-sustaining property of Flowform-treated water was compared with water run through a simple step cascade. A stream of polluted water was split up to run through the parallel cascades, from where it was diverted into separate channel and pond systems.

Comparison of the two parallel streams of water over the four years showed no consistent differences in chemical analysis (van Mansfeld 1986). Temporary fluctuations could not be correlated to the described differences in ecosystem development. Oxygen uptake was similar but nevertheless there was a tendency towards more efficiency in the Flowform cascade.

Drop pictures of samples from the Flowform cascade had more pronounced structures than those from the step cascade (see details of this method on p. 144 and note 17).

Plant growth and macrofauna in water treated with the two cascades was clearly distinct. As seen above, the Flowform cascade stimulated more *generative* development, with more vigorous flower and seed production. The water from the step-cascade induced more *vegetative* plant growth by stimulating leaf development, similar to growth in a calm and shaded, eutrophic downstream area of a river system. Therefore water that has gone through the rhythmical vortical movements in Flowforms appears to lead to ecologically more vital conditions.

The final research evaluation includes the comment that when water is to be used for supporting higher organisms as for drinking water, it would be worth incorporating Flowforms in the preparation system. Also as a polishing treatment for water being reintegrated into the natural water cycle.

Fuller trials would be needed to establish correct type and places for Flowform use.

The installation was transferred into new ownership when the Warmonderhof Farm School moved to another location in the mid-eighties. It is believed the installation is still in use but no more research has been possible.

Biodynamic food production

Much interest has been shown over the years in the use of Flow-forms for water treatment related to food production processes. Particularly in the context of biodynamic food production, the quality of the water used during the process is critical to the quality of the final product. Here, many apparently beneficial effects of Flowforms have been noted by users and researchers. A number of current projects requiring higher commercial output as well as improved quality, are discussed later in Chapter 13.

Seed germination

In studies carried out later at Emerson College, Sussex, experiments with seed germination indicate that plants treated with Flowform water had longer and more regularly grown roots compared to controls grown with water aerated instead with an aquarium pump, where the root length did not differ significantly from untreated water. The biggest differences in root length between the two treatments occurred if the plants were sown at new moon, at full moon these differences were the smallest (Schikorr 1990, Nelson 1987a and 1987b).

In other studies, the rate of germination of wheat was demonstrated to increase by 11% in Flowform treated water compared to untreated water (Hoesch *et al.* 1992).

The forms within dried drops of water on glass slides were distinct: drops from Flowform treated water which had been used to germinate grain, formed pointed, crosslike $CaCO_3$ crystals. Drops from the aerated controls — likewise used to germinate grain — formed ring-like forms around centres, producing a rather amorphous impression (Schikorr 1990).

Figure 10.9. The Laboratory Stackable Flowform here shown in the cellar studio as a complete working unit, with reservoir above and below for treatment of relatively small volume samples. Each Flowform unit consists of the functioning Flowform with a separate supporting piece on top to hold the next form above. It is held together on two wires. Photo: J. Green.

Bread production

Jürgen Strube and Peter Stolz of Kwalis in Fulda, Germany, tested the effect of Flowform treated and untreated water used for baking bread. The treatment consists of running the water over a cascade of granite steps followed by a number of Flowforms. The authors conclude that the amount of water uptake in the bread is higher after treatment and therefore the bread remains fresh and free of mould for at least two more days compared to untreated water. As well as this the volume of dough is increased by 4%, while the taste and consistency of the product is significantly improved, attested by numerous consumers (Strube and Stolz 1999).

Aspects of Flowform use in food production

We are aware of the fact that Flowforms made of reconstituted stone on a cement basis can change the pH of the water flowing through them for a limited period through leaching (Strube and Stolz 1999). Although after a while influence is minimal or non-existent, it is certainly advantageous if Flowforms used for treating drinking water in food processing, are made of ceramic, glass or granite. This is now being undertaken by Tegut of Fulda, Germany, using ceramic Flowforms designed for their purpose.

Other aspects of Flowform treated water have appeared through research studies. For instance, Flowforms, especially those relating to path-curve surfaces, appear to encourage the imprinting of metallic salts which can be present in coloured glazes, into the microstructure of the water under treatment. This is equivalent to a potentizing process and makes the materials more readily available to the organism. These vibration patterns within the water are critical to healing processes (Hall 1997).

Research into quality

As discussed above, an important aspect of the work on Flowforms has always been that of demonstrating verifiable qualitative effects on water after it has passed through specific rhythmical treatment. Here, the usual scientific approach requires a statistically based proof to indicate that some kind of repeatable result is achieved.

On the results of such a kind of 'proof' a method of inquiry is normally deemed useful or not. Its further development therefore depends on a quantitative answer which has a positive or negative finality about it. However, with statistics almost anything can be conclusively 'proved' depending upon how an experiment is set up. Thus statistics might be set up to prove different things from the same conditions.

Where we are dealing with water, life and rhythms, fluctuations in results are bound to occur, and taking into account water's sensitivity and its mediating function one can hardly expect that 'results' can ever be exactly repeated. Nothing in our environment can in fact be exactly repeated — the cosmos or total environment is always changing, albeit within a rhythmical 'order.'

To demonstrate qualitative changes in water which has been passed through Flowforms, methods are therefore needed which extend nature's processes and show up influences more specifically not only in the treated water itself, but also in plants and other organisms exposed to the treated water. For instance, how does treated water affect germination, plant growth and the metamorphic development within a plant? Are leaf and blossoming processes changed in emphasis and what does this signify? How are regenerative and reproductive properties affected?

To investigate such results, a number of 'picture-building' methods are available which, although questioned by some, begin nevertheless to show a direction in which we have to search. Actual changes in the way organisms respond to water can be observed, for instance, those activities in a plant which are more or less strongly influenced by different water treatments. The keeping qualities of nutrient substances are a useful indicator, with important implications for the storage qualities of vegetables.

Fluid samples can be tested through capillary dynamolysis, chromatography, crystallization or the drop picture methods. The first three methods have been practised since the 1920s, and the last since the 1960s. The various test methods used comparatively can together give a wider understanding of the spectrum of effects, a methodology practised for instance by Dr Balzer-Graf with great success.

Capillary dynamolysis

A cylinder of standard filter paper is placed vertically in a special circular dish which contains the sample to be tested. This liquid sample is drawn up through the filter paper over a period of time by capillary action and is then allowed to dry out. This is followed to a higher level by an indicator with which it reacts, at the same time creating a 'picture' which through comparisons can be 'read'. Organic and inorganic materials can be investigated.

Chromatography

A similar method employing capillary action but now using a horizontal disc in the centre of which is an absorbent 'stem' of rolled filter paper. The resultant concentric rings of deposited material can give an indication of the condition of, for instance, soil samples. As with all test methods the investigator has to learn by experience to interpret the results.

Crystallization

Samples are combined with copper chloride which is used as carrying agent. Over a period of some twenty-four hours the liquid is allowed to crystallize out in a special dish placed in a temperature and humidity controlled cabinet. The resulting crystalline formations can again by comparisons be interpreted.

Drop picture method

In a specially prepared polished glass dish the liquid sample under investigation is held. The viscosity of this sample is increased by the addition of a fixed percentage of glycerine. Into this dish, drops of distilled water are allowed to fall from a correct height and regulated intervals. The resulting movements within the dish can be observed by means of an optical system which increases the intensity of the shadow effects created by the difference in viscosity.

The method uses the very movement of water as an indicator of its capacity to carry or mediate information at any given time. [18]

<h1 style="text-align:center">11</h1>

∞

Flowform related developments

There have been a number of developments which, although they have arisen through work with the Flowform method, cannot themselves be called Flowforms. While they do have a relationship to the vortex or even the lemniscate, they depend however upon some kind of mechanical support such as rotation or rocking. This is a fundamental difference to the Flowform as this functions through a dependence upon proportion and gradient only.

Virbela Screw

From the very onset of the work, the question of water circulation was of paramount interest. How could this be achieved without influencing any subtle effect that might be gained by means of rhythm and surface? Though cumbersome in comparison to an efficient motorized pump, the Archimedean screw is ideal for our purposes with respect to the benign way in which it will raise water.

The first consideration must be with respect to the material in which the screw is made and what influence this might have. Toxic materials such as glass reinforced polyester may be convenient but have an unacceptable effect as a substance, and therefore may not be used. Colleagues in New Zealand for instance have used stainless steel units which are available on the market. This it seems is an acceptable solution.

While acknowledging the relative neutrality of the screw function, the next question that arises is whether one could use the process of lifting to contribute something positively beneficial. Could the screw be so designed that in combination it would augment the anticipated effects of the rhythmical treatment provided by the Flowform?

For me it is an ongoing inquiry how, through human ingenuity and a deepening understanding of nature's processes, technologies which could be envisaged in nature's terms, might well enhance her func-

tions. After harvesting crops, for instance, how can this produce be nurtured holistically rather than be dismembered and reassembled in unnatural ways? Although one lacks by a long way adequate capacity to carry this out, attempts have to be made.

With the above in mind, I envisaged a development based on the Archimedean screw. This apparatus consists of an open channel wrapped round what would be a rotating axis. The channel itself is formed in spirals, one turning inwards towards the central axis, the other outwards away from the central axis. As the whole is rotated, water is scooped up at the lower end and falls through the channel as it is carried upwards. It follows a complex spatial movement that is something like an elongated spiralling lemniscate as indicated in the flow diagram (Fig. 11.1). Of course the water moves generally upwards beneath the central axis in the section of the channel, which is at any time at the lowest position with respect to the overall spiral.

One way of simplifying the design of the channel would be to produce two screws each with a channel providing left- or right-handed spirals just falling towards the axis, which is simpler to achieve. Two screws would then be needed, which would enable the water in time to move in both modes, as indicated in the drawing (Fig. 11.2).

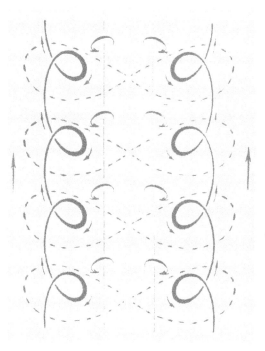

Figure 11.1 (far left). A drawing (050776) intended to convey an impression of the expanding and contracting water movement as it falls through alternate vortices of the channel in the lifting process.

Figure 11.2 (left). A simplified alternative in which the channel would either wind in left- or right handed spirals while each would rotate in the opposite direction. It is namely easier to arrange for the water to fall towards the central axis rather than away from it.

The unit (Fig. 11.3) comprising a section with the left- and right-handed spiral, would be repeated to produce the length needed. There could even be two or three such channels arranged round one axis, interlocking and thus possibly self-supporting. This would be used in combination with a Flowform cascade as intimated above. Such a total unit would be very compact and provide a rich movement treatment for the water.

At the time this was being developed we were almost commissioned to carry out the project in Switzerland. It would have been mounted in an interior architectural setting and be fabricated in a transparent material to provide a fascinating visual experience.

In order to come a little closer to understanding and demonstrating an idea of the path that water would take in such a feature, Herbert Dreiseitl who was studying with us for a period during the late 1970s, made a very inspiring model which turned out to be one expression of a three-dimensional path-of-vortices (Fig. 11.4). Due to the alternating vortices the path combined to provide two symmetrical aspects, the one with vortices turning inwards, the other turning outwards. In order to understand the form it is necessary to observe carefully and slowly a number of times the path taken by this twisting surface.

Such an installation which I have called a Virbela Screw, would also provide an ideal opportunity to incorporate path-curve surfaces. These are discussed later in the Appendix on scientific aspects. Surfaces of this kind are an expression in archetypal mathematical form, of those generated by living organisms to which they are intimately related. From a geometrical standpoint it can be postulated that if water were to be encouraged to follow the asymptotic curves on such path-curve surfaces a quality of balance would be achieved in the nature of the movement between space and counterspace. In order to investigate the effect of such movements upon water and subsequently the water's effect upon the organisms it supports, means of this kind are continually being sought that would allow water to 'absorb' this kind of information.

Figure 11.3 (above top). The section arrived at in collaboration with Don Ratcliff who was studying in the sculpture course at that time. This unit would be repeated to produce a channel of prefered length. The next stage consists of a simplification process which would enable three such channels to be built interlocking round one axis.

Figure 11.4 (above). A model made by Herbert Dreiseitl, later Associate in Germany, to convey an impression of how the water moves in such a so-called Virbela screw. This makes visible a three dimensional path-of-vortices.

Wheel Flow Unit

Another attempt to investigate a means of treating water over a longer period consisted of a unit rather like a tyre, in the hollow interior of which containing surfaces could be built. At rest this shaped channel would hold a quantity of water to the horizontal brim at the inner circumference of the channel. On rotation the water would fall through the channel following the surfaces offered (Fig. 11.5). These surfaces either empirically or mathematically designed induce left- and right- handed vortices in the water as it falls. While in motion, water can be added until full capacity is reached, in fact a circulation can be arranged in that water flows in and out continuously at a preferred flow-rate.

Although such systems as this have been built and shown to function, it has unfortunately not been possible due to lack of resources, to investigate their efficacy in achieving a quality improvement in water. This work we anticipate will become possible now we have the improved facilities in the new Virbela Institute building.

Path-curve Suction Pump

During the 1960s while producing mathematical surface models in Herrischried, I proposed the construction of a tube based on the vane shown in Fig. 11.6. This was eventually made in ceramic (Fig.

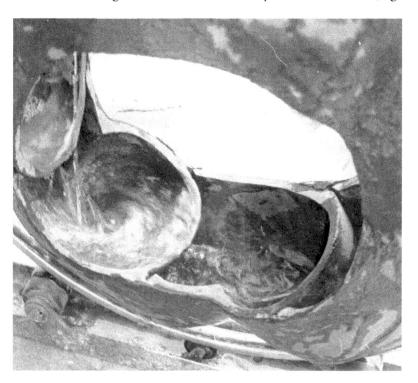

Figure 11.5. The Wheel Flow Unit (040380) while in rotation, demonstrating the water falling through left- and right-handed vortices.

FLOWFORMS: THE RHYTHMIC POWER OF WATER

11.7) in order to eliminate as much as possible any effects introduced by a material. Again it has not yet been possible to carry out an exhaustive investigation with this apparatus.

The rotation of this model on its axis, when primed, causes water to be drawn upwards and be discharged from the upper orifice. While functioning thus as a kind of suction pump, it allows water to caress the path-curve surfaces offered, at the same time closely following the asymptotic curves lying on them.

In the early eighties we devised a plan with Nick Thomas to build a cavity between two vortical surfaces within which dividing walls following the asymptotic curves were built. The plaster prototype with an asymptotic curve drawn on it is shown in Fig. 11.8. Such curves were used to form the vanes which lay between the top and bottom surfaces (Fig. 11.9). The vanes are shown here mounted into the lower surface ready to be enclosed by the top vortical surface (Fig. 11.10). The whole is then mounted with a

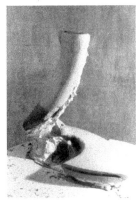

Figure 11.6 (left below) and Figure 11.7 (above).

Figure 11.8 (below). A plaster vortex form (200181) turned by means of a template rotating on a central axis. The Asymptotic curve used as the basis for vanes described at Figure 11.9.

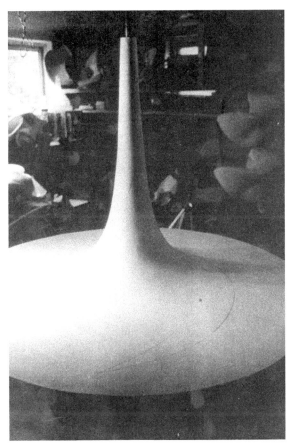

top stabilizing plate connected with a driving mechanism. At the lower end a return valve is mounted to keep the form primed with water (Fig. 11.11) ready to be tested out in the fish ponds at Horsely Mill (Fig. 11.12). Great care has to be taken to avoid electromagnetic fields in the vicinity; and where present, these have to be completely neutralized. Work has still to continue with this apparatus to determine possible effects on the water from its exposure to these mathematical surfaces.

Rocker Flow Unit

The first notions about the utilization of rocking processes concerned the passage of water to and fro through Flowforms between two reservoirs (070870) as described in Fig. 5.27 on page 77.

Figure 11.9 (above). Plan and elevation of the Vortex Pump, built using mathematical Path-Curve surfaces.

Figure 11.10 (below). The interior of the cavity divided by vanes running on the asymptotic curves of the surfaces.

Figure 11.11 (right). The pump mounted to rotate with its lower end under water. The none return valve at the lower end prevents draining once the cavity is primed.

Figure 11.12 (below right). The pump in use over fish breeding ponds, for aeration and simultaneously treatment over the mathematical surfaces.

Figure 11.13. Sketch for rocking Flowforms on steps.

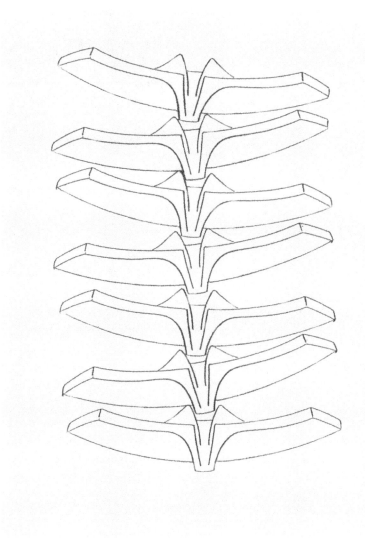

Another idea arose for me from the observation that the swinging process of the water within a Flowform generated energy which could in turn rock the Flowform itself (050774) especially if a curved surface of appropriate design were provided on the underside (Fig. 11.13). It would be quite captivating to observe such a rocking cascade brought into motion by the flow of water. The trick would be to manage the correct tuning of the water movement frequency within the vessels with that of the induced rocking frequency in order that a mutual support and even enhancement might be achieved. Clearly the situation would also

Figure 11.14. A Rocker Flow Unit in use.
Figure 11.15. The same in use with adult and child.
Figure 11.16. The Rocker with the addition of a wooden cradle to enable two people to coordinate their movements and produce optimal lemniscatory movement in the water.

Figure 11.17 a&b (100282) (opposite left and middle). Demonstration of the movement in a Rocker Flow Unit designed for mixing processes. The movement shown is predominantly linear moving in a double lemniscate. It is however readily possible to create a successive process at the central crossing over area. The visibility is enhanced by the use of suspended granules.

have to be clear of earth or granular materials that would obstruct the motion by coming between the moving surfaces.

The idea of a Rocker that Andrew Joiner developed concerned the opposite, namely a rocking motion that generates movement in water contained within the vessel (150281). This enables a child (Fig. 11.14), an adult, or both (Fig. 11.15), to stand on the rim and establish the correct rocking frequency that will generate a double lemniscate in the four-cavity form. Alternatively the cast form can be housed in a wooden cradle which allows two people to experience a coordinating activity (Fig. 11.16). This could also be used for therapeutic purposes.

Two photographs (Fig. 11.17) show a Rocker Flow Unit of about two litres capacity, developed in our studios, which could be used for mixing, for instance in the preparation of medicines. This was investigated but has not yet been taken into use. Only a linear type of movement is represented here as demonstration to show the overall pattern. A strong succussive element can however also be created in the central area.

Such units were also made in ceramic and these can be used, for instance, for rhythmical treatment of fruit juices.

A much larger unit was developed for a capacity of fifty litres (Fig. 11.18). Here it is shown functioning with a thick creamy mix of clay slip. It was also successfully demonstrated with oil and plant material, rather like a thick porridge. In principle the treat-

FLOWFORMS: THE RHYTHMIC POWER OF WATER

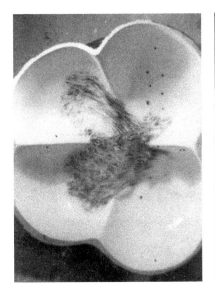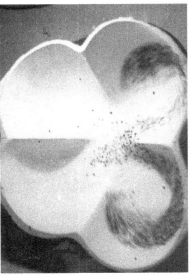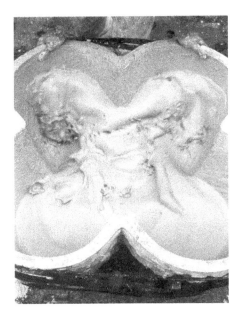

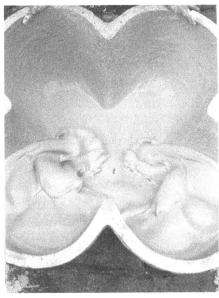

ment consists of resting phases and moving phases over a period of some days. With such a vessel the liquid can remain untouched, and when it has to be moved the rocking motion brings it into a double lemniscatory process. When carried out directly by the operator it is very interesting how the movement can be controlled in a very harmonic way, relating to volume and frequency.

This project, although well on the way and quite successful, unfortunately had to be discontinued when a change of management brought about a sudden withdrawal of the commercial sponsorship for its development.

More workshop photographs are on pages 162–65 with reference to Flowforms around the world.

Figure 11.18 a&b (131083) (above). A larger version of the Rocker, in use with clay slip to demonstrate movement with more viscous fluids. This was also carried out with oil and chopped plant material. This vessel was designed for the purpose of rhythmically treating fluids during extraction processes, to produce therapeutic oils.

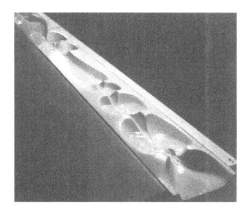

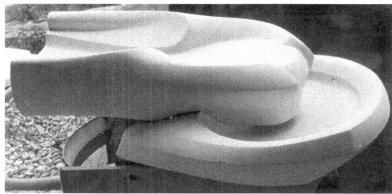

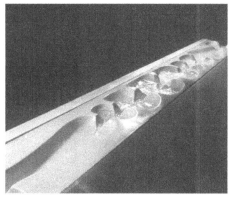

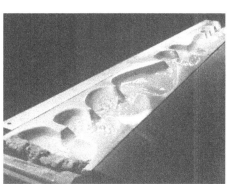

Figure 12.1a–c (above sequence). Plaster proto-type for the Handrail Flowform Cascade at the ING Bank Amsterdam.
Figure 12.2 (above right). The exit form for the first section.
Figure 12.3 (right). A vertical feature as exit form for the last section of the Handrail, contain-ing single cavity Flowforms. These generate rhythmic waterfalls down a two metre drop to the ground floor over the last section of the balustrade.

154

12

The Flowform around the World:
an illustrated survey

From 1982 to 1986, we were involved in a large project for the ING Bank in Amsterdam for which we carried out a number of installations. A new Flowform sequence that I conceived specially for this building was the Handrail Cascade (Fig. 12.1) that was carried out with Nigel Wells. It consisted of two sections. We had entry and exit on the first run (prototype Fig. 12.2; final execution Fig. 12.4) and the final feature on the second run was a two-metre high pulsing waterfall consisting of single cavity Flowforms (Fig. 12.3). Unfortunately as the design was later taken out of our hands and altered, the final result was not as we had intended. The material for the reproduction process of the Handrail is called cold-bronze and was very well carried out by Empire Stone Ltd. of Narborough who put in an extraordinary amount of effort to do a good job.

For all the other projects in the Bank we used Flowforms that already existed: the single cavity at the Directors' entrance; the Ashdown in a number of places including a roof garden; while in two garden areas the asymmetrical Olympia Flowforms and the complete Olympia Flowform complex as described on pp. 125–27 were used (see Fig 9.11b, p. 127).

On the following pages are illustrations showing Flowforms in their various forms and uses around the world.

Figure 12.4. The Handrail Flowform Cascade, end of the first run, at the ING Bank, Amsterdam.

Figure 12.5 (top left). Hanne Keis, Denmark: Open office area by Diax Telecommunications, Bang & Olufsen. This installation is very effective in cleaning up the atmosphere which is loaded with electromagnetic influences from computers. The difference is very noticeable when the installation has to be turned off for maintenance.

Figure 12.6 (top right). Akalla Flowform Cascade (UK) set up in Järna in 1976 for the Biological Sewage system.

Figure 12.7 (above left). Hanne Keis, Denmark: Silent Flow is a cascade with integrated plinths for interior or exterior use. With changing flow-rate the acoustic effect can be varied according to need.

Figure 12.8 (above). Akalla Flowform Cascade (UK) in Steinen in conjunction with the paddling pool of a municipal swimming pool near Basel, Switzerland (executed by Hansjoerg Palm, Arch. Raeck).

Figure 12.9 (left). Amsterdam Flowform (UK) set up as permanent feature for the Floriade Gartenschau Amsterdam, in 1982.

Figure 12.10 (top left). Atelier Dreiseitl, Germany: Landesgarten-schau Reutlingen in 1984 pulsing waterfalls.

Figure 12.11 (top right). Reimar von Bonin, Switzerland: a collaboration with the Flow Design Research Group through the introduction of a Flowform in the central form, thus generating rhythmical waterfalls. Installed for the Weleda in Arlesheim, Switzerland.

Figure 12.12 (above left). Atelier Dreiseitl, Germany: Flowform cascade integrated in steps leading to the market place in Hattersheim near Frankfurth. Made in natural stone, 1990.

Figure 12.13 (above right). Emerson Flowform (UK) in the Garden of Ton Alberts in Amsterdam.

Figure 12.14 (right). Olympia Flowform Cascade (UK) for the 'Duft und Tast Garten' (Scent and Touch Garden) Bundesgarten-schau in Ulm, 1979. Planned and installed by Atelier Dreiseitl. Notice the pulsing waterfall.

Figure 12.15a&b (left above and below) Iain Trousdell N.Z.: A city feature in Wellington built in 1984. Reproduction of Olympia Flowforms originating in the UK.

Figure 12.16 (above top) Iain Trousdell N.Z.: A feature at Manners Mall pedestrianised town centre.

Figure 12.17 (above) Iain Trousdell N.Z.: A major innovative installation stradling the railway in the centre of Hastings, called Haukunui (meaning Large Spring).

Figure 12.18a&b (opposite left above and below) Nigel Wells Sweden: The Stensund Flowform used at Klocagården an Old Peoples Home in Järna, built 1992.

Figure 12.19a&b (opposite right above and below) Mark Baxter Australia: A competition winning installation for the Deakin University, Burwood Campus in Melbourne. 1998.

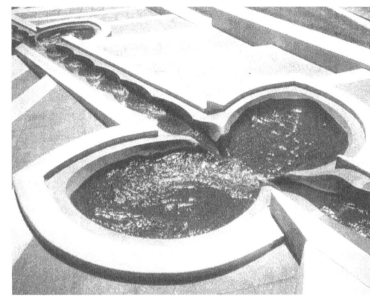

Figure 12.20 (opposite left top). Iain Trousdell N.Z.: A garden cascade at Motucka.

Figure 12.21 (opposite left middle). Philip Kilner UK: Single Cavity Flowforms in beaten copper in an atrium at the Brompton Heart Hospital, London. 1991.

Figure 12.22 (opposite left below). Michael Monzies, France: A single Flowform called 'Papillon' used in an interior installation.

Figure 12.23 (opposite right top). Axel Ewald U.K.: A wrought copper Flowform at The Mount, Tunbridge Wells.

Figure 12.24 (opposite left below). Andrew Joiner U.K: The Rose Fountain in Middlesborough

Figure 12.25a & b (left top and middle). Andrew Joiner U.K.: Several Flowforms from Iris Water and Design at the Hampton Court Horticultural Exchange, 1995

Figure 12.26 (below left). Paul van Dijk Netherlands: The ING Single Cavity Flowform used for many ING Bank offices all over Holland.

Figure 12. 27a & b (below). Stem Flowform Model (UK) originally designed for the Ethnographic Museum in Stockholm. Radial orientation from above and from the side, one of the many ways in which this cascade can be planned.

Figure 12.28 (above). The Wellhead Stackable Flowform, a small unit with which one can build a screen wall. Cascades can be interspersed with separating units. Plants could also be grown in the wall.

Figure 12.29 (right). Some of the products made at Vidaraasen in Norway.

Figure 12.30 (below). The same Wellhead Stackable Flowforms can be stacked vertically in different ways. They were originally designed to be built round the top of a cylindrical wall of a well shaft for treating the water in air and light.

For a water treatment project at Hogganvik Landsby on the west coast of Norway we made a Wellhead Stackable Cascade prototype. This was conceived to form a cylinder tower, an interesting direction that led to the idea of a screen wall in which any number of cascades could be used but also with intermediate elements (Fig. 12.28). This design we realized would have to be carried out in ceramic to be useful for water treatment. Such Flowforms can be used in different ways for instance as a column in a partition wall which has a rectangular aperture to create an acoustic effect. They can be built up with a kind of dual rhythm or be regularly placed (Fig. 12.30).

In May 1989 we started talking with Joachim Eble and Klaus Sonnenmoser about the Ökokultureller Gewerbehof in Frankfurt. The interior site consisted of a ramp thirty metres long and seven metres wide as entrance area between ground floor and first floor. This was situated within a five-storey glasshouse providing circulation pathways and staircases as well as main features of the air-conditioning system, namely streaming water and plants. Nick Weidmann worked out the complicated levels with wheel-chair path, 'express-path' with steps and the critical cascade gradients. We made the entry or 'Quellform' with a single cavity Flowform, sending rhythmical waves over a large surface along the edge of which a pulsing waterfall develops (Fig. 12.31). Large shallow Flowforms followed with the intention of bringing stone slabs over

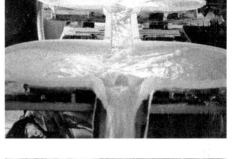

the edges in connection with planting, thus the level rims (Fig. 12.32). Smaller asymmetrical Flowforms came in the next section (Fig. 12.33). For a north glasshouse we designed a unit 80 cm x 40 cm x 40 cm to be duplicated at various levels along a wall. The water issues over an edge and moves in waves over a vertical wall and with many hanging plants forms a humidification feature (Fig. 12.34).

From 1991 onwards we began projects for Herrmannsdorfer Landwerkstätten at Glonn near Munich, in conjunction with a curing-cellar air-conditioning system. There was interest in working with various rhythm frequencies, so we needed a simple Flowform solution which could be adjustable dependent upon the results of running experiments. This was achieved by creating units that would make up a Flowform but between which cut bricks could be

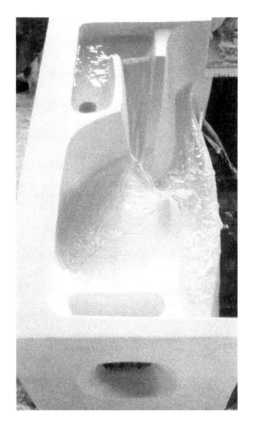

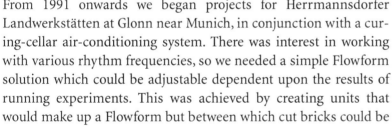

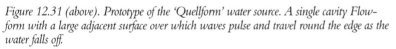

Figure 12.31 (above). Prototype of the 'Quellform' water source. A single cavity Flowform with a large adjacent surface over which waves pulse and travel round the edge as the water falls off.

Figure 12.32 (above right). Again for the same project a large flat Flowform in sequence, designed to be integrated in a stone surface punctuated with plant pockets.

Figure 12.33 (above middle). Still lower down an asymmetrical Flowform sequence to be built within a plant bed integrated with stone surfaces.

Figure 12.34 (right). In a north glasshouse of the Gewerbehof the Wall Flowform, to be built in two blocks of Flowforms with vertical wall extending below them over which water pulses. In conjunction with generous planting this creates a humidification unit.

Figure 12.35 (above left). Glonn II prototype for half the flowrate at 30 lt/min.

Figure 12.36 (above middle). The prototype of Glonn I designed to generate vigorous rhythms in connection with water treatments. This can also be used for decorative purposes in a landscape mounted on a straight wall as shown in this demonstration or on walls which are curved. The Flowforms can be swivelled through 180° to form any curve or a vertical tower.

Figure 12.37 (above right). Glonn II. This is produced in stoneware ceramic and can be hung vertically on stainless steel wires. It can be used for instance within vats in which fruit juices are circulated.

Figure 12.38 (opposite right top). Prototype for a single cavity Flowform to be assembled in a seven-sided complex for treatment of water discharged into a cattle drinking trough.

Figure 12.39 (opposite left). Cattle drinking trough at Herrmannsdorf with two cascades mounted and in operation, surrounded by animals.

Figure 12.40 (opposite right middle). Prototype of a path-curve Flowform under construction being tested for function with water.

Figure 12.41 (opposite bottom right). A later stage in the prototype production. This Flowform incorporates surfaces related mathematically to the vortex and to the spiral of a cow-horn. It is destined for investigations in connection with organic Biodynamic liquid farm preparations.

inserted to enlarge the cavity and thus reduce the frequency of the rhythm. These were inserted into the connecting tunnel between the Herten cascade in the shaft and the actual curing cellars.

During this period we also worked on a new Flowform which was eventually called the Glonn I as it was used there first. This Flowform has a number of possibilities as it can be swivelled through 180° in plan. It can be stacked vertically or mounted for instance on a straight or curved wall (Fig. 12.36). Glonn II is a slightly simpler version with half the flowrate (Fig. 12.35). This is now being made in ceramic and can be hung for instance on wires (Fig. 12.37).

The last of some seven projects for the Herrmannsdorfer Landwerkstätten was the Cattle Drinking Trough (Fig. 12.39), a seven-sided complex with two single cavity cascades and trough; here is shown the unit prototype in preparation (Fig. 12.38).

In 1993 we began with Nick Thomas to build a special Flowform incorporating path-curve surfaces (see Appendix 3) for the purpose of mixing biodynamic preparations. These path-curve surfaces were designed to relate mathematically to the surface of a water vortex together with that of the cow-horn. The reason for this is to relate the rhythmical mixing process, on the one hand to the vorti-

cal movements of the water, and on the other to the substance being mixed, which in some cases is the specially prepared cow-dung. However even if other substances are to be mixed, a relation to the cow is not foreign to the farming process. In any event the apparatus is intended for research purposes, not as a final solution, and needs investigating further.

The surface was constructed in four sections, left- and right-handed, forming the quarter sections of the Flowform, here being tested with circulating water for correct function (Fig. 12.40). We expect to see an enhanced effect by combining rhythm with surfaces related organically to the plant and animal substances being mixed. The prototype is shown here as far as we have come to date (Fig. 12.41).

13

Present and future

New work with Flowforms

Today's consumers are becoming increasingly sophisticated and demanding in the matter of quality — perhaps above all in food production but also in health and cosmetic products. Quality-sensitive products are in demand and there is a subsequent need for companies to deliver higher output. We find more and more enterprises wishing to investigate how Flowforms can deliver commercial levels of output within the preparations and treatments that contribute to their products. We know from long experience in biodynamic farming that it is possible to incorporate Flowforms into large volume operations, but each new application demands its own study and solution. To show the wide variety of new directions under study or development, some current areas of commercial interest in our work are summarized below.

Water transport and treatment

We have discussed water's capacity to absorb even aggressively from its surroundings what it does not possess within itself in the form of energy or mineral content. General questions therefore arise about the effects of transporting water via pipes, pumps, tankers and even bottles and how these processes affect its quality. Can ways be found to treat water effectively so it still contains the vitality needed when it reaches the end user? This is a complex problem depending on many factors.

Absorption of the kind described above can be disadvantageous to health. For instance, there are healing spring waters which may no longer have their optimal qualities when they reach the patient. With the need to take in water on the one hand, it can be problematic when minerals and vitality are lost by excretion of the very water that is intended for support. We are addressing this question with a spa in Italy known for its healing properties. Here the water has to be pumped, transported by tanker and pipe and finally stored in a reservoir before it is used. Can the water still have its total vital qualities and if not how might these be enhanced again by passing the flow through a nature friendly rhythmical lemniscatory treatment? These questions are being examined and constitute fundamental research for the Virbela Rhythm Research Institute (see p. 170).

Food processing

In connection with food processing, water quality is increasingly a factor considered by producers. It is reasonably well established in the the context of bakery production for instance, even if mainly anecdotally, that Flowform-treated water affects the rising of the dough, the texture and taste of bread as well as its keeping qualities. Such effects await confirmation and optimization in collaboration with research institutes that have already carried out rigorous testing. Meanwhile one organic bakery where small glass Flowforms are being used is at the Herrmannsdorfer Landwerkstäetten south of Munich.

In Germany again, Tegut with its main production factory at Fulda, (see p. 142) produces Demeter label meat, milk and bakery products for its supermarket chain within half a day's delivery. For the water needed in the bakery a novel treatment using a particular granite is carried out in combination with Flowforms. Further installations with ceramic Flowforms are under construction.

For Bio-Sophia of Lillehammer in Norway we are contributing Flowform technology to various aspects in the production of grain-milk: washing, swelling, cooking, fermenting and diluting. Demand for grain-milk has risen greatly because of increased lactose intolerance in the Norwegian population, and there is a large and growing export market for the products to Germany under the Demeter label.

Hermann Dettwiler of Inverta AG, a Swiss production line designer, is the main planner. The new central five metre high water treatment plant consists of three ceramic Flowform cascades, with a 6,000 litre storage reservoir below, and will be housed in appropriate projected polyhedra which I have designed using George Adams' geometrical methods. Wide-ranging research indicates that biological events taking place within such spaces are positively

Figure 13.1. From the Metalwork Studio David Fuchs, Deggenhausen, the completed 'Flowform Prism' water treatment installation was transported to Lillehammer, Norway and mounted for Bio Sophia in May 2003. It consists of three ceramic Flowform Cascades with spiralling path-curve channels, which direct the treated water into the 6000 lt. tank below. The protecting glass and steel structure above consists of a projected Icosahedron, the reservoir a projected Dodecahedron both expressing as dual polyhedra the gesture of the Golden Mean. The central column through which water enters, supports the cascades which hang on wires. Water continues circulating by means of a spiral Mono pump until needed in all aspects of the grain-milk production process for grain-milk from a number of different cereals. The first stage consists of washing the grain which is carried out by passing it through large Vortex Flowforms.

influenced, and we intend to investigate this thoroughly. Our growth tests are already indicating enhanced effects. It is especially intended that an artistic element is reflected in the design of the production unit, the whole being conceived as an organism offering a qualitative continuity of those nature friendly processes involved before the harvest and which need to be further nurtured to enhance the nutrient value of the products. The design and nature of the processing path as well as the consciousness of the personnel involved, are all considered to be of paramount importance.

From 2004 with Biodynamic apple producing Firm Augustin near Hamburg a washing and sorting installation functions in conjunction with a Sevenfold Flowform cascade built by Peter Müller which promises also to improve storage. For some time we have been involved with treatments in preparation of fruit juices. In 1980, we provided an installation for Kaare Godager in Norway, whose production unit was taken over eventually by Ole Vestergaard who now makes products from biodynamic harvests in Norway. He will increasingly be treating water for his firm, Corona Safteri, using ceramic Flowforms. We are preparing to continue the development of a large rocker vessel for Corona in Trondheim that will enable a volume to be moved through a double lemniscate by a rocking motion.

Already in 1974 I was working with Flowforms which rocked on a curved base by virtue of the swinging water movements themselves. This led to the building of a large rocker that enabled water to stream back and forth between two reservoirs through a two-way cascade thus avoiding the need for transportation. In 1981 Andrew Joiner made a four-leaf clover shaped vessel in our workshop that moved its content by mechanical rocking. It was used for play and therapy. This idea, taken further during the 1980s for treating liquids but temporarily abandoned for organizational reasons, is now to be rejuvenated for our client Corona.

Health products

Our work with rhythms is underpinned by the original research of the German chemist Dr Rudolf Hauschka whom I was privileged to meet in 1960. Wala, the pharmaceutical company that he founded near Stuttgart, has had long experience of working with rhythm and water. Hauschka set up an experiment in the 1920s to reveal the preservative effects of rhythm, within the framework of searching for new ways to manufacture medicines. He put rose petals in two vessels filled with water, one of which was exposed to certain rhythms. The petals in the untreated water rotted in a matter of days, during which time the petals in the treated vessel dissolved to a beautiful ruby red solution with no signs of corruption. The resultant aromatic extract had a stable life of over thirty years, without having recourse to alcohol, the traditional preservative.

Wala now make natural extracts and essences for medicinal and cosmetic purposes, using some 150 different herbs and plants cultivated on their own land. We are expecting to follow up further involvement with Wala in the treatment of water and fruit juices for which they already have ceramic Flowforms. The company is interested in rose-water and rose-oil production, where it wishes to study the possible uses of Flowforms for irrigation water treatment. In the longer term, studies may be set up to examine the use of Flowforms for large-scale water treatment in the production of medicines. This direction indicates a main theme for future research of our Institute.

In preparing water specifically to irrigate plants, such as roses, the buds reveal specific mathematical surfaces the numerical values of which can be transferred to the water as information. Such insights we intend to develop in order to optimize the use of water in its support of the growing plants and their subsequent processing. This question of mathematical surfaces is discussed more fully elsewhere (see Appendix 3).

Looking to the future

Global concerns

In November 2001, the United States Department of Agriculture reported a shortfall in world grain harvest of 54 million tons. In the previous year the shortfall was 34 million tons. As a result, grain reserves have slumped to 22 per cent of annual consumption, which is the lowest level for twenty years.

Drought was, and continues to be, a major factor in this pattern as in conventional modern agriculture it takes a thousand tons of water to produce a single ton of grain. But in many food-producing regions of the world, water tables are falling, or urban needs are increasingly depriving agriculture of essential water supplies. An Earth Policy Institute review of this problem comments that:

The adequacy of food and water supplies are closely linked. Some 70 percent of all water that is pumped from underground or diverted from rivers is used to produce food, while 20 percent is used by industry and 10 percent goes to residential uses. With 40 percent of the world's grain harvest produced on irrigated land, anything that reduces the irrigation water supply reduces the food supply.[18]

The balance between water supply and food supply is therefore critical, not only socially and economically but politically, too. In this context, it must be seen that any treatment that enhances the fertilizing and life-giving power of water must have the potential for important applications. Available research and experience suggest that the Flowform could contribute to such an enhancement.

The Virbela Rhythm Research Institute

We see the establishment of the Virbela Rhythm Research Institute in Sussex as a vital step forward in exploring the potential contribution of Flowforms to these vital water issues of global concern.

The effort to set up such an institute has had a long and difficult history going back over twenty years. With building work at last going ahead in 2002, we are on the point of establishing a properly designed centre that will be concerned with rhythm research essentially in connection with water. It will take some time to embed the new centre as an organic entity into its surrounding environment, one of our aims being to create an appropriate water landscape around the centre. In terms of global outreach, we hope that the Institute can develop further as a focal point and a catalyst for useful ideas. A strong networking connection has been active for some time already, bonded by those who have come from all corners of the globe to study or research here at Emerson College.

The principal aims of the Institute will be as follows:
— carrying out work calling for an enhanced consciousness of the true nature of water and of its function as mediator between rhythms of environment and organism; rhythm being here considered as a basic necessity for life, essentially carried by watery media;

— carrying out scientific research on the influence of rhythmical and metamorphic processes; demonstrating subtle water movement phenomena and observing natural phenomena;

— researching into movement and surface; investigating both empirically and mathematically designed surfaces;

— undertaking design research related to the implementation of rhythmical processes, including Flowform design research;

— educating towards a wider social consciousness concerning the sustainable use of water; offering short courses and conferences on specific aspects, involving guest speakers;

— promoting a better understanding of rhythm in all realms, artistic, scientific, cultural;

— consultation, offering advice and professional expertise dealing with some of the many practical problems relating to water and the environment.

— promoting work with Flowforms internationally.*

Some areas of future research

The Virbela Screw development (see Chapter 11) has in recent times been taken up again through the fact that Wilhelm Nickol of Helixor Stiftung and Fischermühle Landgut has agreed to sponsor a completion. So work is progressing in collaboration with Georg Sonder. An interim metal unit has been prepared, the design will be prototyped and possibly produced in conjunction with Oslo University where computer equipment will enable us to simplify and incorporate preferred mathematical surfaces. The intention with this screw is to positively treat the water while lifting it, after which it can flow down a cascade of Flowforms. This enables a very compact unit to be produced for special water treatments in connection, for instance with food-processing.

From the 1960s when I was producing mathematical models for George Adams, there is a good deal of material that can still be used for research. One unit is in ceramic and forms a suction pump which we have an idea to investigate within an egg-shaped vessel we expect to borrow in the near future. This will combine various

---

* A full description of the aims and constitution of the Institute is in Appendix 4.

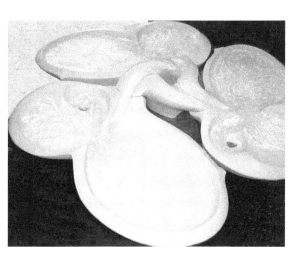

Figure 13.2. Plaster prototype of Viktors Flow-form in our workshop.

mathematical surfaces the effect of which we can test. Another larger existing version of the vortex suction pump from the 1980s we intend to continue investigating.

Apart from these areas of application we will continue with our basic research on the complex aspects of rhythmical processes within Flowforms and also attempt to penetrate the secrets of the subtle proportions which lead to them. This is connected with the study of organisms and the way in which rhythms are generated everywhere in the living world.

One of our present tasks will be the design of a new dynamic radial Flowform probably something like three metres across. It is likely, in order to make its use more manageable, that we shall divide it into three sections with an extra central inlet feature (Fig. 13.2).These will naturally be linked together but remain separate units. This project arises from the wish of Henry Nold not only to support our Institute with a significant contribution, but to celebrate the work of the Austrian physicist and inventor Viktor Schauberger. Although I knew nothing of his work when I started developing the Flowform method, I became aware during the 1970s of the universal importance of his contribution to our understanding of the nature of water. He was one of the early pioneers who drew attention to the influences of modern science upon the natural world, a comprehension of which increasingly demands our active participation. Our activities are certainly compatible, and to such a degree that some have assumed that Flowforms are also an inspiration of Schauberger's, an assumption about which I certainly feel honoured.

A summary

Our blue Planet supports abundant life through the fact that it has a water cycle fed by its oceans extending over most of the surface and brought into movement by the Sun. The conditions seem to be unique in the known Universe but the search is on for evidence of any other such bodies. Water dissolves and transports substances and influences, coarse and subtle. It is an element that sacrifices itself entirely to its surroundings. It comes under the influence of gravity and levity and due to these polar opposite agencies, it moves. Life moves within it and it moves within living creatures and these movements are always rhythmical. Life forms always have to do with surfaces or skins which convolute to create shapes and bodies. It is such surfaces which exist within the volume of water whenever it moves and which act as organs of mediation for all the formative processes relevant to the sustaining of life.

Movement and surface

Through controlled and protected laboratory work, it is possible to watch the unfolding of veil-like surfaces within water, and even more sensitively within air. It is essential to investigate the reproducibility of these sensitive phenomena and the influence upon them of diurnal, seasonal and planetary rhythms.

Designed surfaces

The seminal work carried out during the 1940s and 1950s by George Adams and Olive Whicher at the Goethean Science Foundation in England and concurrently by Theodor Schwenk at the Weleda laboratory in Germany, provides the basis for the Virbela Rhythm Research Institute's further activities.[20] Since then the continuous investigations of Lawrence Edwards and supportive research of Nick Thomas have confirmed the legitimacy of those initiatives taken to investigate the influence of specific surfaces upon water. Through the work of others, eminently within the realm of potentization, it is increasingly well known that water carries information which is independent of substance.

Since the inception of the Flowform Method in 1970 at the Institut für Strömungswissenschaften, it has become possible not only to investigate the influence of rhythms upon water but also inseparably that of the surfaces essential to the carrying of such rhythms. These surfaces can either be empirically or mathematically designed. As already discussed, the relationship to life-forms of so-called path-curve surfaces was discovered by George Adams. We are able to incorporate such surfaces within the design of Flowforms allowing water to caress them intimately. The investigation of these surfaces in relationship to water constitutes a very exciting field of research and constitutes one of the most important themes for us at the present time.

From my early work in this area, 1970, it was immediately demonstrable that the Flowform method could be utilized as a tool for the investigation of influences upon water of path-curve surfaces, as well as the space and counterspace concepts of the vortex.

Not only is it a question of testing their effects by incorporating these surfaces within the design of Flowforms, but also of investigating the surfaces and curves which water itself manifests when induced to move in various ways. The primary task is to discover how the proportional relationships of all parameters involved can bring about an optimization of the life-sustaining qualities of water. How can the sensitivity of water to the surrounding conditions be so enhanced that organisms are truly embedded and supported within their environment?

Rhythms in water

Water, which represents all fluid processes, is the carrier of rhythms. However in outer nature, that is on land and sea wherever water moves, rhythms are continually appearing and disappearing due to the multitude of ever changing influences, only within the organism is a rhythmic stability achieved. Rhythms are essential to life and there is no life without rhythm. The Flowform method demonstrates the fact that rhythm is generated through a quite precisely proportioned resistance to a fluid momentum of controlled intensity in any given set of circumstances. Rhythm is itself a gateway, a medium by means of which life can flourish. It becomes manifest physically as the phenomenon of metamorphosis. Metamorphosis and enhancement are key features of all life processes.

The path-of-vortices generated by a straight-line movement through still water demonstrates water's potential for order and metamorphosis. It is this phenomenon that inspired the idea of building an organ of metamorphosis for water which in turn led to the Flowform itself.

Consciousness of water

Generally speaking, consciousness of the spiritual dimension of life has been much diminished during the last centuries in order for humanity to learn intensively about our physical surroundings. One fundamental aspect that is interwoven with everything expressed in this book is our acknowledgment of the living spirit that is present behind all the physical phenomena we have discussed. Approaching our task from this perspective, together with a new understanding that the Earth is a living organism, as reflected in recent exploration of Gaia theory, all adds support to the idea that a healthy and harmonious revitalizing of nature can take place only through awareness of the spirit.

Further, as a closing comment, I would like to say loud and clear that it is a vitally important social task today to work on the idea that water is our most precious resource for which humankind needs to develop a universal and alert consciousness. It is a sad and sorry reflection on international politics that we hear the cliché so often uttered that 'the wars of the future will be over water.' If cooperation could be achieved at international level over the issue of the quality of water, tensions arising from water shortages worldwide could be lessened, with human energy being directed towards a better understanding that could produce solutions, rather than towards potential conflict.

As we have tried to make clear throughout this book, the water cycle has maintained itself as a life-giving resource for millennia by means of its ability to move water through many different states. These states have to do with constant movement and even more specifically with rhythmical movement, together with a quality of surface that is intimately connected with life processes. In a word, all organisms are connected and associated through water.

Our urgent task must be to learn to understand and work with such parameters in order to enhance the original functions of water in nature with which we in the modern world are so strongly interfering. The next generation must develop a completely new attitude to water if survival in many arid parts of the world is to be guaranteed, and indeed the same necessity applies to maintenance of the quality of life in industrialized nations.

Appendix 1

Metamorphosis*

What is metamorphosis?

What is metamorphosis? First of all we can safely say this many-faceted phenomenon is one of nature's ways to stimulate in the human being a mental and spiritual mobility. It has to do with diverse and changing relationships, between or among elements of any given totality or organism, associated usually by dynamic processes which are not of a physical nature. A familiar example would be that of a plant, with its root, cotyledons, leaves, blossom, stamen, seed and fruit, the relationships of which already exhibit different types or qualities of metamorphosis.

It is important to note right from the start of these considerations that a distinction must be made between those processes of change that are physically discontinuous and those that are physically continuous. The former may be called metamorphic, the latter, growth processes. Although there is an element of continuity as for instance with the stem of a plant which holds it all together, each organ of the plant is separated by a physical gap. Each leaf is separated from the next and, finally, also the blossom. But each leaf goes through a change of form as it grows and this is quite another kind of change to that shown between each successive leaf (Fig 14.1). Likewise in anatomical terms Fig. 14.2 shows vertebral bones in metamorphosis (above) and growth (below). This difference can also be observed within geometrical processes where the origin or archetype of the form remains the same for a growth process, whereas for the metamorphic process the archetype changes (see Rotational metamorphosis, p. 184).

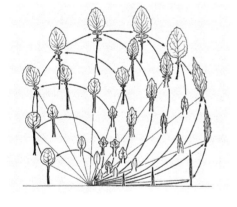

Figure 14.1. The outer ring shows the mature leaves of one plant, left (bottom) to right (top) as metamorphosis, a physically discontinuous process. The spiral lines show the form stages through which each leaf passes during its physically continuous growth process in the opposite direction. The radiating lines link these stages, with the representative outer mature leaf sequence. From Jochen Bockemühl, Erscheinungsformen des Ätherischen.

---

* Introductory parts of this section were originally written in collaboration with Mark Riegner of Prescott University, U.S.A.

Rudolf Steiner further elaborated the methods of observation initiated by Goethe (see Chapter 3). In his books *A Theory of Knowledge Implicit in Goethe's World Conception* and *The Philosophy of Spiritual Activity*, Steiner built a philosophical, yet practical, foundation for developing the latent possibilities of cognition explored by his predecessor. Moreover, Steiner maintained that if one could school oneself to apprehend the inherent principles manifesting in, and giving order to, natural phenomena, one could then create forms artistically which would express an inward consistency. The results of such an application would be no naturalism — a copying of nature — but a free expression whose activity would be based on an intuitive grasp of formative principles operating in Nature.

Steiner applied this approach in various artistic endeavours, perhaps most notably in the architectural design of the two Goetheanum buildings in Dornach, Switzerland, the first of which was destroyed by fire in 1922. In the original auditorium he designed a series of seven wooden columns each with a capital and base of unique form. The seven columns formed a metamorphic sequence which gave expression to an internal wholeness, a unified gesture. Thus the individual columns were related by a common ideal element not unlike the way plant species are related to each other by their ideal archetype (see also *Metamorphosis* by Frits Julius).

In this initial example, perhaps one of the most profound and universal of metamorphic phenomena is elaborated. Such a sevenfold process is indeed fundamental to any evolving process that takes place in time and is rarely to be found manifest in an actual spatial sequence. It was with this sevenfold sequence that Steiner began his artistic work in 1907.

This little observed and little understood realm of metamorphosis now became the foundation for a new organic architectural impulse within which all the Arts can be embodied. To be truly organic, architecture must above all evolve out of metamorphic principles. In the first Goetheanum building Steiner used also a number of other metamorphic relationships, to be discussed below. (For more on the Goetheanum buildings, see for instance *Der Bau* by Karl Kemper and *Eloquent Concrete* by Rex Raab.)

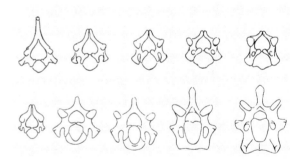

Figure 14.2. These are preparations I made from cows. Above are the separate cervical bones of a cow embryo from left to right, the 7th, 6th, 5th, 4th, 3rd. These constitute a metamorphic physically discontinuous process.

Below five drawings of the physically continuous growth process starting with the embryo and moving through time shots ending with the adult. These are preparations of the 6th cervical bone from five animals, indicating how this bone grows. This drawing is unique and shows the results of research not otherwise published. It shows the equivalent anatomically to Bockemühl (opposite) botanically.

Beginning a study of metamorphosis

Metamorphosis has to do with relationships of which there are as already stated manifold qualities or types. The word means 'change of form.' This is usually from one object to another and can happen in many different ways. A general definition concerns the relationship between two or among a sequence of physically separate forms within an organic whole. Metamorphosis indicates either a very wide-ranging concept or it might be limited to much more rare examples. My feeling is we should risk the wider, freer view because we need today to find access, especially artistically, to its manifold implementation, that is, to artistic and scientific forms that embody a relationship between physical and spiritual dynamics. Today we are all too bound by our comprehension of phenomena through an emphasis on the material, physical aspects of existence. The manifold nature of metamorphic processes can become evident through the characterization of a number of metamorphic types and this follows below.

Metamorphic types

These qualities of metamorphic type have been defined and experienced in courses starting in the mid-1960s at Emerson College, Sussex; and Rudolf Steiner Högskolan in Järna, Sweden. A more detailed description may later form the subject of a separate book.

Rather than limiting my attention by concentrating on the sevenfold sequence (see Chapter 3) as a means of studying metamorphosis, I was inspired to seek for myself ways to penetrate the bewildering array of such phenomena in nature, in order eventually to come to Art forms which would themselves reawaken an experience of the relationships between spirit and matter.

The task we shall undertake is to describe through a study of the phenomena, examples of different types of metamorphosis such as polar, development, evolutionary, and so on. This is intended to bring clarity to the contemplation of a realm that is so complex. We need such a key to enable us to penetrate with creative activity this dynamic aspect of morphology.

My entry into this area of experience came through meeting

Ernst Lehrs and eventually through knowing his book *Man or Matter* in which an example of leaf metamorphosis was given. I began collecting leaf specimens simply because this was a most convenient method of beginning a study of such processes, bearing in mind that an intensive examination of one realm can reveal universal secrets. Although comprising only part of the plant organism the leaf is a fundamental organ out of which, as we have seen with Goethe, all other organs metamorphose.

> The green leaf is the child of the blue heavens and the yellow sun.
> It emerges from the dark earth and heralds the colourful light of the blossom while providing us with our harmonious green-carpeted planet.

An organ of transition, the leaf exists between the linear root and the planar blossom (see Adams & Whicher, *The Plant between Sun and Earth*). The leaf stem maintains an albeit dwindling memory of the root whereas successive forms reach towards the blossom which is anticipated by the leaf blade.

Description of metamorphic types

It seems necessary in order to enhance our understanding of metamorphosis and other transitional processes, to describe how forms relate to each other and what quality of relationship they have. Metamorphic relationships occur namely among the various component parts of a total organism and not essentially within a single evolving part, although this also takes place.

Is it adequate to think in terms of one possible 'true metamorphosis'? Rather than limit ourselves as already indicated, to any one very special case it seems best to develop a discussion on a broader basis. Nature is indeed penetrated with many types of metamorphic process. Nevertheless in order to become aware of them we have to sharpen our consciousness and observation.

The following descriptions constitute an attempt to explain a limited number of different types. It is not a question of becoming fixed in this description of types but of continually rediscovering them for oneself as distinct. I am sure that through our individual approach to the matter other types will certainly be formulated:

Polar (root and blossom)

Polarity generates movement, because a balance between the two opposite tendencies has constantly to be re-established. Every physical object, for instance with its degree of hardness or softness, depends upon its specific relationship to polarity. This 'movement' leads to rhythm, as attempts are repeated to find equilibrium. Metamorphosis is a physical manifestation of rhythm relating variously to time and space. Mankind and Nature exist between opposites which are manifest in many forms; expansion and contraction; plane and point; planar and linear; concave and convex; hot and cold; day and night; where these polar phenomena can be experienced in terms of metamorphosis.

Metamorphosis must also be allowed to take place in us through an observation of phenomena. We cannot expect root to change into blossom for instance, but we must experience the metamorphosis within ourselves.

Figure 14.3. Sow thistle, this gives a very fine example of Development metamorphosis, a step by step change from a spread in the blade width away from the plant stem, to an emphasis in width next to the plant stem.

180

Development

This is demonstrated by an even, sequential change between opposite (polar) conditions such as the leaf process between root and blossom. Leaf forms develop step by step without significant change in the plant substance. Here is illustrated a Sow-thistle with spiral phylotaxis (Fig. 14.3).

The *Wirbelstrasse* (path of vortices) also demonstrates a development metamorphosis and is in a sense archetypal (see Chapter 4 p. 49ff).

Sevenfold

This exclusively sevenfold process has its own specific laws and has to do fundamentally with an evolving situation in time, whether it be a social organism, a living organism or a planetary organism. We have for instance the characteristics of a first stage, a second stage, a third stage in any such evolving process. The middle state is followed by a kind of mirroring or echoing of the first three with a final achievement at a higher level in the seventh. For the first time this theme was expressed artistically as a spatial phenomenon by Rudolf Steiner in 1907. As a natural phenomenon the sevenfold is an expression of a time process, as we experience in the seven year periods in human life or in the days of the week. In both of these we have a connection with the planetary worlds, as the names of days of the week imply. There we have a microcosmic reiteration of the great evolutionary processes of the earth (for further detail, see Steiner's *An Outline of Esoteric Science*).

Vertebral

The sevenfold phenomenon is hardly ever expressed within physical nature, as a sequence in space, one exception however is in the vertebral column. The phases are named from the tail to the head: caudal, sacral, lumbar, dorsal, cervical, axis/atlas, and cranial; this sevenfold process demonstrates a hidden reference to its evolutionary origin. (see pp. 92–3). Organisms evolve embryologically from a head which grows a 'tail.' Thus it stands as a type progressing from a development (between opposites of head and tail) to a phased evolutionary metamorphosis.

In mammals and human the neck bones are reportedly also always to be seven in number, nevertheless the axis and atlas are intimately

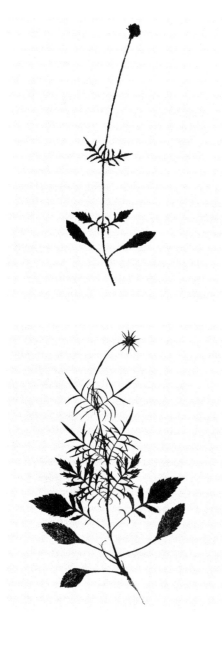

related to each other and to specific functions and are thus distinct in their shapes and basically unlike the other five cervical bones.

As the vertebral column originates within the universally human and is manifest throughout the kingdom of mammals, as a metamorphosis of gesture it is entitled to a name of its own. It does however at the same time exhibit the traits not only of development and a sevenfold evolution, but also functional and polar metamorphosis. It is such an important and universally manifest type that it warrants a detailed discussion elsewhere.

Planetary

Metals, woods, colour: here also a sevenfold relationship is involved. This type of metamorphic process takes place within the observer as an inner experience. Metals do not for instance change into each other but our experience of them changes us.

The following provide examples: lead (Saturn), gold (Sun), silver (Moon), iron (Mars), mercury (Mercury), tin (Jupiter), copper (Venus). In the same order: Hornbeam, ash, cherry, oak, elm, sycamore, birch.

It is good to remind ourselves here about the work of Lawrence Edwards in *The Vortex of Life* where he shows that these plant species actually respond in their fluctuating shape to the positions of the planets to which they traditionally belong (see also Georg Schmidt on the influence of planetary constellations on the growth of related tree species).

Position

This includes a number of different qualities of metamorphic process, three of which follow:

Location

From an original form, plan or motif a metamorphosis can take place in a specific direction or to a particular location, for instance details in an architectural setting. From the plant world an example: a field of buttercups standing side by side show similar variations on the theme of this species, when however the buttercup appears away from the area in continually changing ground conditions for instance up a slope, the manifestation of the species can change even significantly. In such circumstances there is no exact demarcation where a metamorphic quality of relationship begins. Different species of the same

Figure 14.4. Two Scabiosa plants found in different locations showing a metamorphic change in the form.

182

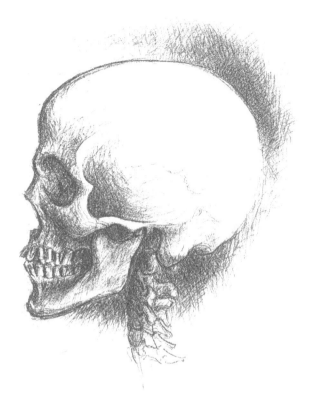

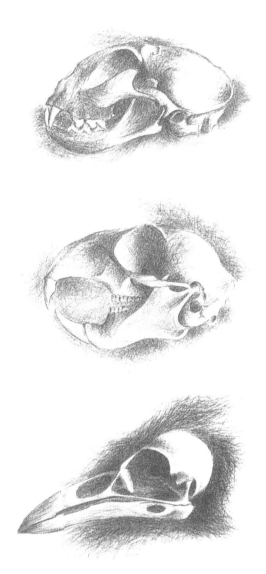

family clearly show metamorphic relationships but this would have to do with Gesture. One can observe this in the Scabiosa for instance.

Gesture

From the human as origin, the animal forms are metamorphoses, a particular aspect of the human is expressed in every animal species. The human is the unifying principle or idea, in Goethe's sense, from which all animal forms arise. Aspects of the nerve-sense system appear as Rodent, the metabolic system as Herbivore and the rhythmic system as Predator. These are the major groups among which many combinations can be found. The human (Fig. 14.8) is compared with the predator, (Fig. 14.5) the rodent (Fig. 14.6) and a bird (14.7).

Function

A change of form often accompanies a change of function, as the bones within an organism can dramatically change their shape. Here are the cranial bones of the vertebral column, the occipital, the first Sphenoid, the second Sphenoid and the Vomer. Even more dramatic is the change of caterpillar to butterfly.

Figure 14.5. Animal forms are one sided expressions of the human. The animal shows for instance a specific capacity and emphasizes it right into the skeletal build. Here the cat, representing the Predator with large canines.
Figure 14.6. A squirrel, representing the Rodent with an emphasis on the incisors (no eye teeth)
Figure 14.7. A bird, the same bone elements, now emphasizing the beak.
Figure 14.8. The human skull from which the animal forms are metamorphoses of Gesture. (drawings 14.3-6 by Axel Ewald)

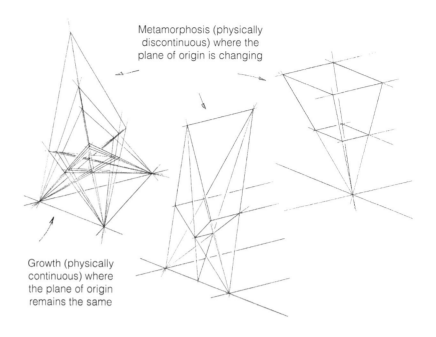

Metamorphosis (physically discontinuous) where the plane of origin is changing

Growth (physically continuous) where the plane of origin remains the same

Figure 14.9. Three projection types of the cube; (left to right,) from one finite point, from two finite points, and from three finite points. The fourth type is the regular cube projected from the points at infinity. A rotation process of the figures brings a change in the plane of origin from which they are projected, thus we call this a Rotational metamorphosis. Drawing: Wilkes.

Rotation

By projection from a finite line or plane of origin, a two or three dimensional form can be drawn. When this is rotated the points and lines within the plane or line of origin, move. From each of the new positions, constantly changing projections are generated. For instance when a cube is projected from three finite points in a plane, its rotation as a three dimensional figure changes the orientation of points within the two dimensional plane of origin. From these points, the changing projections emanate. In other words the three dimensional form must withdraw from space into the plane, in which changes then take place, after which new three dimensional forms can arise. (This contrasts with a growth process which is indicated when a family of forms is created by projection from the same plane or line of origin.)

Random

This is a process taking place in a design context where a number of different shaped forms are used in random order. The forms themselves are metamorphoses of each other but are not necessarily used in a specific order, as in the example of the Akalla project (see Fig. 8.18, p. 120 and Fig. 8.26, p. 122). This type would probably not be found in the context of nature.

Art Form

A particular motif, say that of the Madonna and Child, where one form creates a protective gesture for another, can find expression in different visual art forms, in architecture, in painting, in sculpture. This holds good also for the performing Arts, drama, music, speech, eurhythmy.

Reincarnation

In another sense this is the process indicated by the metamorphosis of the first Goetheanum building to the second. The first building presented primarily an interior space based on two cylinders and domes, a larger and a smaller. The second building was designed primarily as an externalization of the first, a turning inside out.

Cyclic

Second and succeeding years of leaf development in a perennial plant. The annual presents a development metamorphosis which appears during a subsequent year in a different form (e.g. hellebores) and finally proceeds into the blossom.

Plant groups

Frits Julius describes in his book *Metamorphosis* the seven groups of plants. As a Goetheanist he observes plants in their relationship to the vertical and the horizontal, to contraction and expansion:

First three groups: dominance of vertical with subordination of horizontal; dominance of vertical with support of the horizontal; proliferation in the vertical.

Second three groups: reverse of vertical and horizontal.

The seventh group: relates to a balance between vertical and horizontal usually found in annual plants. Respectively (examples): Saturn (conifer), Jupiter (deciduous), Mars (bush), Mercury (creeper), Moon (cactus), Venus (alpine), Sun (annual). Julius continues by explaining these characteristics in relation to stages in human life.

Figure 14.10 & 11. The two Goetheanum Buildings by Rudolf Steiner. He designs the first (top) as an interior architecture out of necessity, to create a space in which he could generate specific activities. This inner content he then externalises by design in the second building (bottom). The first is for a limited group of people, the second, headquarters of a world society. The first in wood the second in cast concrete. A turning inside out as well as a hardening and stabilising process. A kind of reincarnation of an idea. (drawing Axel Ewald).

Appendix 2

Flowform types, designs and applications

A. Overall range of Flowform types

The existing scope of the Flowform method can be seen from the following list which records the range of types conceived by the author. Each type is identified by its date reference number, with an accompanying description of its features and functions.

These designs are copyright and are to be used in collaboration with the author or his representatives. The Flow Design Research Group can give advice and specifications on a selection of the following ideas and on their availability.

The symbols below indicate various categories of development: those types which have been investigated, some of which are in regular use as production models; those for which a design has been made but which have not been put into production; a number of designs developed only as scale models; and others not marked that are still to be investigated.

* Investigated and in use + Designed but not in production   ‡ Scale model only

FLOWFORMS: THE RHYTHMIC POWER OF WATER

1. Double cavity, symmetrical *

This is the kind of Flowform derived from the first experiments when symmetry was being considered in contrast to water's tendency towards asymmetry. (010470)

2. Double cavity, asymmetrical *

It was clear from the first moment that the rhythm was the result of resistance and not symmetry. (research notes 8, 10 April 1970)

3. Single cavity *

This arose from experimentation with asymmetrical forms as one of the lateral cavities is gradually reduced in size, finally to disappear. (research notes 8 April 1970) (210183)

4. Vortex *

A double cavity Flowform with holes in the base which maintain strongly pulsing vortices. (research notes 10 April 1970) (011193)

5. Element

Flowforms can be made up by using block elements which when combined on a sloping surface (like a weir) form cavities which break up the flow into a multitude of pulsing movements. (research notes 30 July 1970)

6. Radial *

Usually designed with three Flowforms in one unit fed from below through a central entry. (original notes 1 August 1970). Original model (151179) developed large scale (200881). Later designs, some with incorporated elevated entry, achieved notably by Nigel Wells, Iain Trousdell, Herbert Dreiseitl, Andrew Joiner and Thomas Hoffman.

7. Path-curve +

Incorporation of mathematical surfaces within the Flowform to investigate their qualitative effect upon water, these surfaces are related intimately to organ building processes. (300770) (020292)

8. Metamorphic sequence *

A number of Flowforms of changing shape and size within a sequence. (240770) Different types of example: Akalla (121175), Olympia (290776), Herten (040578), Sevenfold (051185 & 200686), Hiram (010697).

9. Two direction +

A Flowform built to accommodate both directions of flow, for use on a cradle with reservoir vessels at both ends to obviate the necessity and influence of pumping, intended for tests on qualitative effects of rhythmic treatment. (070870)

10. Tube +

Sections of tube can be used with appropriate chord length removed, mounted on a flat surface with the correct aperture (290371). First four-cavity conceived in this manner (see 12/030471 below).

11. Return *

A Flowform without forward opening but an inner and outer ring between which the oscillating water is returned to an exit in the rear (080471). Later investigated by Andrew Joiner.

12. Four cavity +
This was examined first of all with the Section (or tube) Flowforms cut from a pipe. (030471).

13. Step ≠
A block into which an appropriate cavity is cut, used as a segment of spiralling or straight steps so that a channel is formed through which water can cascade rhythmically. (020471) (040697)

14. Complex
A Flowform with large and small cavities based on the idea of atrial and ventricular cavities of the heart. (100474)

15. Dome *
Translucent or transparent Flowforms which can be used at high level to be viewed from below thus allowing observation of water movements against the light. (Acrylic 290874)

16. Three cavity *
Two inlets with a central cavity mediating between the two outer cavities. (research notes 04 Sept 1974) (220180), (second edition HD 81)

17. Multi-stream +
Exit divided into three channels to utilize pulse issuing alternately to left and right. (101074)

18. Flood control +
In a series of large Flowforms excess water is retained as the forms fill up by virtue of the oscillation. As the flood water subsides the system empties out having provided a buffer. (research notes 1975)

19. Changing rhythm +
At different levels within the Flowform there are distinctly changing diameters to accommodate changes in flow-rate; as the flow-rate increases the Flowform fills up and converts to a slower rhythm (020676).

20. Transport +
The Virbela Screw is a unit consisting of an open spiralling channel wrapped round an axis which when rotated raises water through alternately left and right rotating vortices. (050776).

21. Stem ≠
Each Flowform incorporates a stem through which water enters. This holds the form aloft enabling a pulsating waterfall. (161077)

22. Fish ladder
Fish are attracted by a rhythmical stream of water issuing from a Flowform cascade. This fact could be used to guide fish to ladders which cannot always be built in locations where they are readily found by the fish. (research notes 1977)

23. Regulation: Stream or river ≠
An aperture of specific proportion, in the form of two large blocks or shaped islands, presented as a conical funnel to the flow built on a relatively wide smooth base can generate large swinging wave movements in the stream flowing through. This might be used in an area where bathing or paddling could be permitted. No energy input is required. By increase in flow-rate at times of flooding, the installation

becomes submerged presenting no untoward problems but may also serve to reduce dangerous momentum (research notes Sept 1978). Investigated for the large city flood canal in Vienna.

24. Five cavity combination of three and two cavity Flowforms ∓

A scale model was made as a student project for a paddling pool complex about 4 x 6 m, using a natural stream of several thousand litres/min. (research notes 1978)

25. Floating

Flowforms on buoyancy raft moored on a river to allow water to flow through to generate rhythms. (research notes Sept 1978)

26. Sphere

A Flowform blown in glass against a mould to form the inside surface, surrounded by enclosed air space to the blow-pipe. (research notes Sept 1978)

27. Massage

A large Flowform in which the patient could even lie or rather be slung, so that pulsing water passes under the body in close proximity. Alternatively Flowforms can be used to prepare water prior to water massage in a bath. (research notes Sept 1978)

28. Covered *

Sealed treatment channels to contain gases or curtail loss through evaporation. The water surface must be free to oscillate. In principle used for instance in tropical areas where large-leaf plants can be used for shade and condensation of water or in conjunction with fruit-juice treat-

ment where oxidation needs to be prevented. (research notes Sept 1978) (011180) (190581) (111296)

29. Rotary vessel

Two series of Flowforms enclosed within a vessel, on top of each other and facing, reservoir at each end. To be rotated on a lateral axis enabling a given volume to be treated over a longer period without pumping. (230778).

30. Irrigation

Flowforms used to transport water to irrigation site while treating rhythmically. They can be perforated in the base as required, to allow for seepage directly onto the land until empty at the lower end of the cascade. (070179). A modified Järna could be used for this purpose.

31. Circumference

A ring of Flowforms at the edge of a circular disc mounted at an appropriate angle with a pivot in the centre on which it would rotate. The whole disc would roll round the centre allowing a contained volume of water to fall through the peripheral channel. (210279)

32. Embankment ∓

Flowforms which relate to units designed to insert into almost vertical embankments often used for landscaping gardens in steep terrain. (see Löffelsteine CH) (research notes 040679).

33. Demonstration channel *

From straight laminar movement to an oscillation generated by proportions of the channel, to a meander form, which gradually transforms to a symmetrical Flowform (230679). As idea it contained all Flowform types, part was carried out.

34. Screen +

Flowforms which can be used to build a screen in conjunction with designed elements which support them. These Flowforms are in fact building blocks shaped to function with water flowing through at a desired gradient. Such screens through which there is a limited visibility can be used to divide patio areas and create some shade, while influencing the micro-climate with cooling water flow and evaporation. (150879)

35. Tower +

Upper and lower surfaces are horizontal so that Flowforms can be stacked on top of each other, accommodating the gradient within the form from entry to exit. (150180 M. Arbrecht) Smaller model (180687).

36. Water-bell

This constitutes an enhancement of the vortex issuing from the Vortex Flowform (research notes 12 Mar 1980) also thought of in conjunction with path-curves.

37. Wheel +

A tyre-like unit with alternating forms creating a channel in it, through which water streams in left- and right-handed vortices as rotation takes place. (040380)

38. Handrail +

Sections of cascade to create continuous run on 12% gradient. (100580)

39. Cylinder cascade +

Asymmetrical Flowform for spiral installation in cylindrical vessel for fruit-juice treatment. (011180)

40. Vertical Screw +

Lifting by rotation and suction over path-curve vortex surfaces with possible rhythmic process in the discharge flow. (200181)

41. Stacking convertible unit *

A unit used in conjunction with the Järna Flowform to allow for vertical stacking or 180° change of direction. (180181)

42. Stackable *

A small laboratory Flowform with steep gradient generating vigorous movement, produced in thermal plastic (190581) to be hung on wires. Later produced in glass (010190), ceramic (250496) and eventually in slip-cast (013097). Other larger models have been prototyped or are already in use.

43. Pendulum +

A form that can be mechanically swung or rocked to generate vortical movements. (A.Joiner 1981)

44. Rocker *

A four-cavity form with a domed base on which it can be rocked to generate a double lemniscatory movement. (A. Joiner 110581)

45. Sling

Vessel containing water can be moved to generate lemniscatory movements round a patient who is carried in a sling close to the pulsing water. Could be particularly apt for small children and babies. (This was discussed in detail with interested practitioner, but financial resources did not materialize). (190685)

46. Adjustable *

A Flowform made up of ring sections fixed to a flat surface. The diameter can be enlarged with extra blocks which can be cut to change rhythm to any frequency (260391) as desired.

47. Wall *

A single cavity unit for stacking against a wall issuing a pulsing curtain down a vertical surface for humidification. (010991)

48. Four cavity vortex (N.Z.) *

Four cavities the two rear ones of which have base exit holes, this water can be led again forward into the stream at the exit. Alternatively the holes can be in the forward cavities.

50. Rocking

A typical Flowform but with a curved base, so arranged that the pendulating movement of water within the vessel (lemniscatory) generates a rocking motion of the whole Flowform. This has to be so organized that the movements are mutually supportive. (050774)

51. Convex *

Elevated entry causing a fast stream which by virtue of correct resistance of a suitable kind generates rhythms over a convex surface swinging to the left and right, thence spilling over the forward edge. The resistance can be created by the change in gradient or even a gentle hollow in which water collects to resist the stream.

52. Vertical single cavity *

These Flowforms can be used to generate a pulsing waterfall (for instance in conjunction with the Handrail Cascade in the ING Bank, Amsterdam).

B. Flowform designs and specification examples

The development of an idea

An attempt has been made to maintain as complete a list as possible of authorized design work in connection with the development of Flowforms which has mainly been at Emerson College. There are however several colleagues to date who have contributed to this work but who have moved or returned to live in different parts of the world after attending courses there. Drawings or photograph are included or referred to elsewhere in the book wherever possible. The list is unfortunately not complete and in some cases without description due to lack of information.

Those marked * have been in use for sale.

Flowform/ date — Conceived/in collaboration with

Original Basic channel/ 010470 — Wilkes
The channel through which the first indication of rhythm was discovered.

Double/100470 — Wilkes
Enlarged unit made up of Flowforms of two sizes in which the first lemniscatory movement became clearly evident together with the influence of one form carried over to the next.

Ceramic sequence/210470 — Wilkes
Used for the first experiments carried out at the Flow Research Institute in Herrischried.

Basic Flowform/010670 — Wilkes
A simple but deeper vessel intended to improve the dynamic of the lemniscatory movement converted later to the Järna Flowform for use in the sewage project 1973.

Sevenfold basic (O)/240770 — Wilkes
A simple model with lead walling on a sheet of aluminium demonstrating the idea of a metamorphic sequence following the original question posed regarding an 'organ of metamorphosis' for water.

Torso Flowform (path-curve)/300770 — Wilkes
A test model evaluating the Flowform Method for the incorporation of 'path-curve' surfaces to investigate their effects upon water (Fig. 5.18, p74).

Section (pipe)/290371 — Wilkes
A geometrically describable Flowform (cylinder sections on a flat surface) used to specify the parameters involved, namely; gradient, relating to: aperture dimensions and distance from entry to exit with overall diameter and flow-rate, where the function generates a pulsing lemniscate. The optimal function occurs in the balance between filling up and emptying out of the system.

Slurry/170773 — Wilkes*
One of the first Flowforms brought into operation for the Järna project. Designed only as an interior vessel for quick and simple execution in glass-reinforced cement.

Sewage*/270773 — Wilkes
A slightly larger modified version of the Slurry
Flowform. Used in Järna project.

Open Flowform*/261273 — Wilkes
A larger flatter Flowform allowing alternate spillage
over front edges. Basis for the Emerson Flowform.

Complex/100474 — Wilkes
Over a period many ideas were considered with a
combination of forms seen as a kind of organ with
a number of different movement patterns.

Emerson*/261273/ 260575 — Wilkes
The first sculpturally designed Flowform modified to
contain water, without spillage over the front edges,
for use in interior exhibition in Stockholm (Fig. 15.1).

Figure 15.1 (below left). Emerson
Figure 15.2 (below right). Acrylic

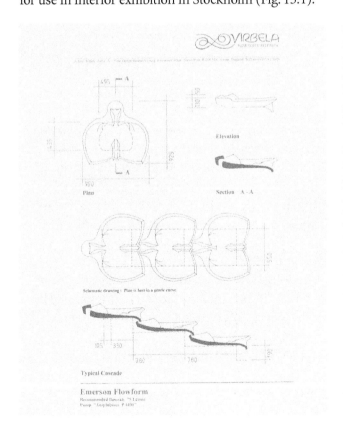

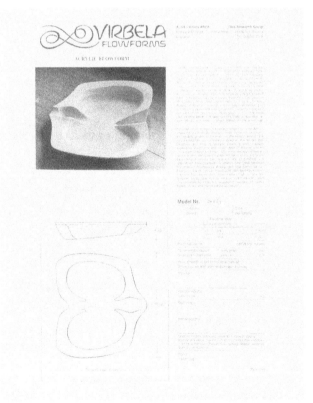

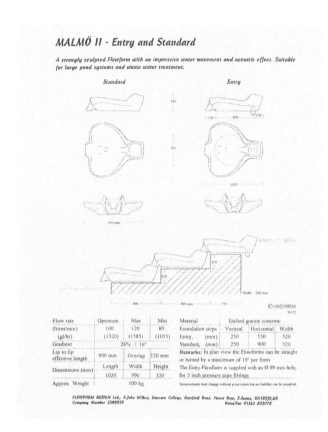

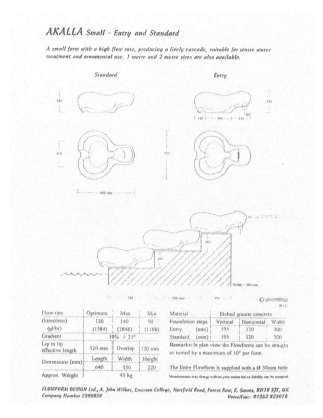

Acrylic*/290874 — Wilkes
For production in transparent or translucent acrylic material by heat and air pressure (Fig. 15.2).

Malmö*/250475/ 071175 — Wilkes/Wells
Increase in size and flow-rate generating dynamic movement for water treatments (Fig.15.3).

Akalla*/121175 — Wilkes/Wells
Flowforms in three sizes for the boulder landscape in Akalla, as metamorphoses of one idea in different sizes to facilitate use on varying gradients. For use in random order (Fig. 15.4).

Virbela screw pump/76 — Wilkes/Ratcliff
A Flowform related project (see Chapter11) based on the Archimedean Screw with an open twisting channel wrapped round an axis, to be used for lifting water while carrying it through a series of left- and right-handed vortical movements thus providing a compatible treatment relating to and servicing a subsequent Flowform cascade.

Brofjord/030576 — Wilkes
A working scale model for large conical Flowforms to be cast in situ, intended for oil refinery (Sweden) or indeed any large-scale oxygenation lagoons.

Järna*/ 010670/300576 — Wilkes/Wells
Developed from the Basic Flowform to replace the original cascade in Järna (Fig. 15.5, opposite left).

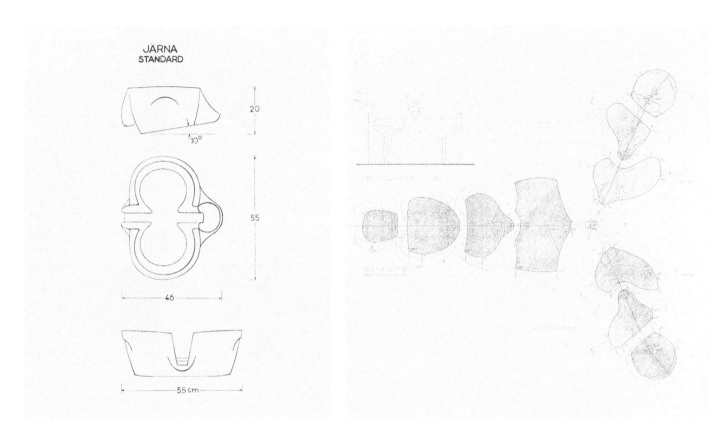

Stacking unit (Järna)*/180181 — Wilkes/Wells
A separate unit used in conjunction with the Järna Flowform to enable vertical stacking or change of cascade direction through 180°.

Olympia (Sundet) (Nyköping)/290776 — Wilkes
Original sketch intended for Nyköping, a symmetrical complex consisting of seven Flowforms, symmetrical and asymmetrical.

Olympia: two thirds sketch/ 010976 — Wilkes/Wells
Preliminary design as preparation for full-size execution.

Olympia: full-size (Fig. 15.6 and 15.6a).

Figure 15.3 Malmö (opposite left).
Figure 15.4 Akalla (opposite right).
Figure 15.5 Järna (above left).

Figure 15.6 Olympia: full-size (above) and Figure 15.6a (right). Figure 15.6a gives the number dedicated to each Flowform design and the date which supports identification.
1) (entry) 200177 — Wilkes/Wells and others
2) 251176
3) 261176
4) 271176
5) 290177
6) large right 170177
7) small left 190177
8) & 11) (exit) 080277
9) large left 130177
10) small right 140177

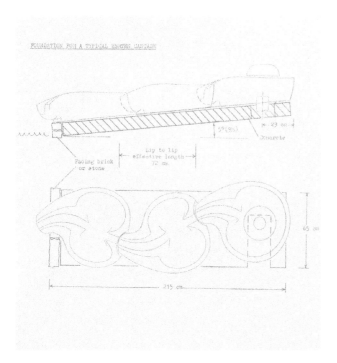

Figure 15.7 (above left). Herten
Figure 15.8 (above right). Herten Flowform
cascade at Michael Hall Kindergarten, with a
crowd of fascinated children.
Figure 15.9 (opposite left). Ashdown I
Figure 15.10 (opposite right). Ashdown II

Glacier (Bonin)*/280877 — Bonin/Wilkes/Trousdell
Incorporation of a Flowform as the middle vessel
of three, at the request of Reimar von Bonin, a
collaborative design.

Ethnographic model (Stem)/161077 — Wilkes
Known as the Stem Flowform intended for execution in metal.

Herten x 2*/040578 — Wilkes/Trousdell
Asymmetrical Flowforms with butt joints
enabling minimal gradient of 9% to be achieved
(Fig. 15.7).

Four cavity*/78 — Wilkes/Trousdell
Two smaller and two larger cavities the rear ones
of which have holes in the base.

FLOWFORMS: THE RHYTHMIC POWER OF WATER

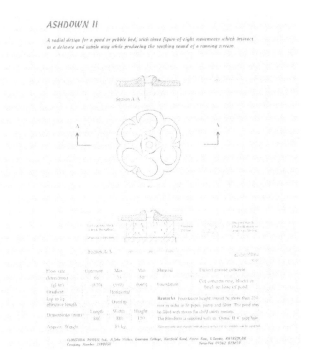

Olympia x 7 (N.Z.)*/79 — Trousdell
Iain Trousdell helped with original design. On returning to New Zealand he repeated the design to overcome financial restraints connected with exporting of moulds.

Large Stackable/250280 — Wilkes/Arbrecht
Horizontal above and below, very flexible for use on a range of gradients, plaster prototype exists.

Demonstration channel*/ 050979 — Wilkes/Baxter
Intended to show a range of different Flowform designs starting with a pulsing stream, through meander to symmetrical and asymmetrical forms.

Ludvika/151179 — Wilkes
Proposal for radial Flowform which became Ashdown and Amsterdam.

Akalla (D)*/80 — Dreiseitl
Amended version of original designs, by stretching the inlets, intended use for varying gradient in the original design, is eliminated.

Three cavity I/220180 — Wilkes/Wiveson/Stolfo
Flowform with two parallel inlets, one central and two outside cavities generating weaving lemniscatory patterns across the whole form.

Three cavity II*/81 — Dreiseitl
Another edition of the same design idea.

Ashdown I*/ 020280 — Wilkes
Originally intended as scale model for larger execution, made into unit with peripheral drainage channel for garden and patio use (Fig. 15.9).

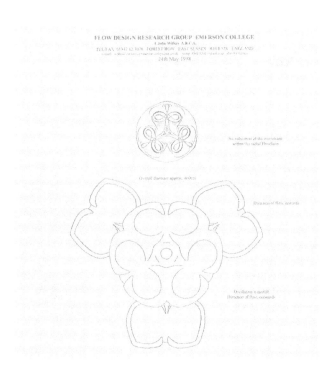

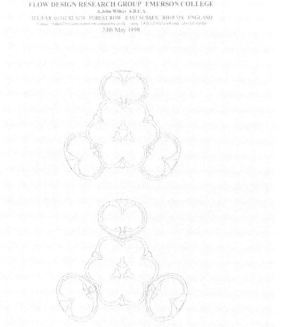

Figure 15.11 (above left). Amsterdam I & 2

Figure 15.12 (above right). Amsterdam I &
Scorlewald

Figure 15.13 (opposite left). Stackable I

Figure 15.14 (opposite right). Rocker

Ashdown II* — Wilkes/Wells
Without peripheral channel for use mounted over pond or boulders. Enlarged and redesigned as Amsterdam I to three metres diameter (Fig. 15.10).

Godager*/011180 — Wilkes
Asymmetrical Flowform designed to fit into a one-metre diameter steel cylinder as spiralling cascade for fruit-juice treatment. Executed in both metal and ceramic.

Handrail NMB*/010580 — Wilkes/Wells
Cascade specially designed for what is now the ING Bank, Amsterdam in conjunction with the handrail, accompanying wheelchair ramp with about 6% gradient.

Amsterdam I*/200881 — Wilkes/Joiner, Wells
A radial design consisting of three Flowforms with central entry originally designed for the Floriade

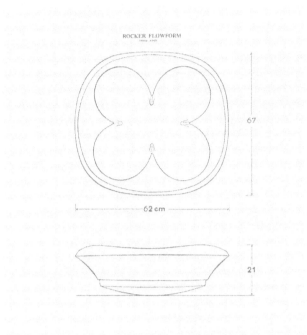

in Holland (Fig. 15.11). Used with the Scorlewald (Fig. 15.12).

Amsterdam II/121081 — Wilkes/Joiner, Wells*
Single large Flowform in conjunction with Amsterdam I, placed at water level in three locations to enable discharge to the centre with rhythmical movement (Fig. 15.11).

Herten x 2(N.Z.)/ 070483 — Trousdell*
Second edition for use in New Zealand.

Gänsersdorf radial/85? — Dreiseitl*
Elegant radial Flowform designed on a stem.

Stackable I (plastic)/190581 — Wilkes*
A small Flowform intended for laboratory use, easily produced in thermal plastic and vertically stackable with a supporting unit in between each Flowform (Fig. 15.13).

Stackable II (plastic)/ 010981 — Wilkes
The same basic design as a double unit held on a central pipe which provides entry for water at the top, each subsequent form rotated through 90°.

Stackable III (glass)/ 010190 — Wilkes*
Redesigned for glass produced as pressing, to be hung vertically on steel wires.

Stackable IV (ceramic) — Wilkes
To mount as cascade in the top of a wall accompanying steps, with entry and exit units.

Rocker (615mm x 665mm)/150281 — Joiner*
Four cavity clover-leaf design with rounded base on which one or two participants can stand and generate lemniscatory water movements through rocking. An extended cradle was designed to accommodate for therapeutic use (Fig. 15.14).

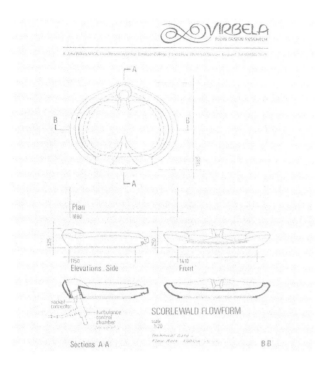

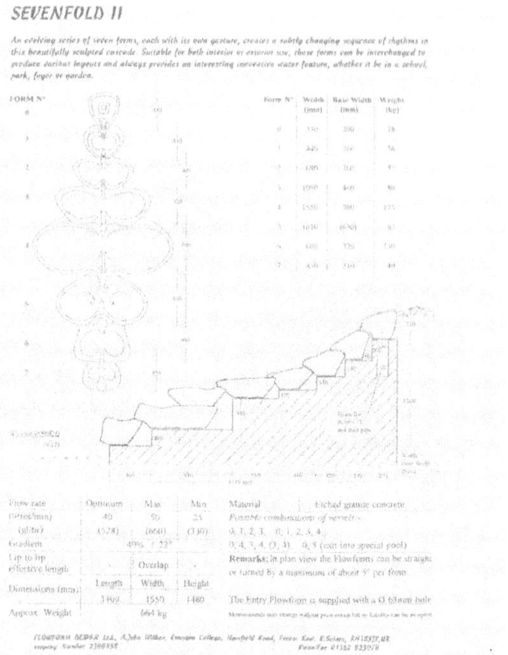

Figure 15.15. Scorlewald
Figure 15.16. Sevenfold II

Scorlewald*/010282 — Wilkes/Joiner, Wells
Derived from a section of the Amsterdam I for combined use, with wide lip across which pendulating waterfalls can develop (Fig. 15.15).

Miroma (Australia)*/ 140682 — Baxter
Ceramic (Australia)*/ 150682 — Baxter
Beehive (N.Z.)*/ 041282 — Trousdell
Victoria (N.Z.)*/ 120584 — Trousdell
Taruna (N.Z.)*/ 140485 — Trousdell

Shire (Single Cavity)*/210183 — Kilner
A single cavity Flowform provided with an extended surface over which wave patterns can develop and over the one edge of which a waterfall oscillates.

Cloverleaf (N.Z.)*/290985 — Trousdell

Large Rocker (Wala)/83 — Wilkes/Monzies
A deep large-volume (50 or 100 lt) rocker commissioned by Wala for pharmaceutical preparations. Plant extracts to be made over a period during which recurring movements can be generated at preferred times (prototype exists). To be used on a cradle.

Plastic Rocker (Wala)*/83 — Wilkes
Deep middle-sized rocker for experimental purposes at Wala. For use on a cradle.

* Sevenfold I x 7*/051185 — Wilkes/Wells
First edition of original idea (demonstrated in Sevenfold 0) made in three pieces with fixed joints and sloping foundation, for economic reasons (later separated out for further development). Used as basis for design of second edition.

* Sevenfold II x 7*/200686 — Wilkes/Palm
Seven separate Flowforms mainly used in sequence with entry (and occasionally exit for interior use). Advantage taken to vary gradient on stepped foundation for small and larger Flowforms. The first three or four Flowforms have frequently been used for smaller cascades. Certain other combinations are possible (Fig. 15.16).

Sevenfold III/010198 — Wilkes/Weidmann
The entry and No. 1 are combined 0/1 and now designed as also with No. 7, as Flowforms without re-entrant surfaces, this assists casting.

Wellhouse stackable*/180687 — Wilkes/Weidmann
Horizontal above and below, asymmetrical, for multi-cascade in cylinder or wall formation with possible alternate none-Flowform support.

Broceliande (derived from Emerson)* — Grégoire
The interior surface of original design maintained but the mass increased, thus losing the intended elegance of the Emerson Flowform.

Silene (derived from Järna)* — Grégoire
A revised edition of the external form of the Järna.

Siloe (derived from Järna)* — Grégoire
A straight block of four Flowforms for practical use but no flexibility in plan.

Morgan (derived from Ashdown)* — Grégoire
An elaborate successful design based on the radial principle.

Butterfly* — Grégoire

Nymphea* — GrégoireDyane (derived from Malmö and reduced)* — Grégoire
An edition of the Malmö design reduced in size carried out with a student.

Single Cavity ING*/88 — van Dijk
Following the NMB (ING) project, Wilkes recommended the idea of a feature specially designed for all the Bank locations in Holland. Paul van Dijk was commissioned.

Hattersheim*/90 — Dreiseitl /Wells
A very successful unique project carried out in granite for the public market square.

Stensund*/90 — Wells
A small composite design for use in exterior and interior projects.

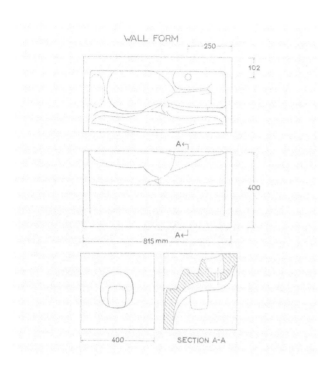

Frankfurt Quellform*/201189 — Wilkes/Weidmann
A single cavity Flowform with waterfall entry, conversion to rhythm followed by large surface over the edge of which water falls, pulsing. Used also in Hamburg Waterworks.

Frankfurt Large*/100490 — Wilkes/Weidmann
Shallow Flowform with flat edge to accommodate surrounding stone surface.

Skanderborg from Frankfurt*/92 — Keis
Based on large Frankfurt Flowform for extensive project in Denmark.

Frankfurt Small*/ 011090 — Wilkes/Weidmann
Asymmetrical Flowform for fixed mounting in gentle curve, used as third stage in Frankfurt interior air-conditioning project.

Wall*/300290 — Wilkes/Weidmann
Rectangular dimensions 80 cm x 40 cm x 40 cm as repetitive unit below which a vertical surface receives pulsing waves. Single-cavity Flowform (Fig. 15.17).

Garden*/92 — Weidmann
Small open Flowform used mainly for cascades in private gardens (Fig. 15.18).

Rockery Ceramic*/ 93 — Wilkes
A small ceramic press Flowform available in different stoneware glazes, for use indoors or out and can be placed on stones or fixed *in situ*. (diam. approx. 23 cm) (Fig.15.20).

Ceramic Flowform (slipcast)/93 — Wilkes   Patio I*
Small Flowform for slip-cast intended for production in handicapped workshops and use in conservatory and interior (Fig. 15.21).

Figure 15.17 (opposite left). Wall
Figure 15.18 (opposite right). Garden
Figure 15.19 (below). Glonn II
Figure 15.20 (right). Rockery ceramic
Figure 15.21 (right middle). Patio ceramic
Figure 15.22 (right below). Glonn I

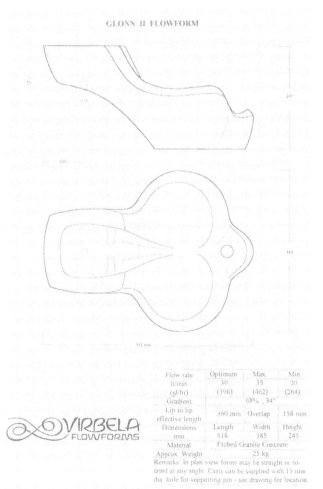

GLONN II FLOWFORM

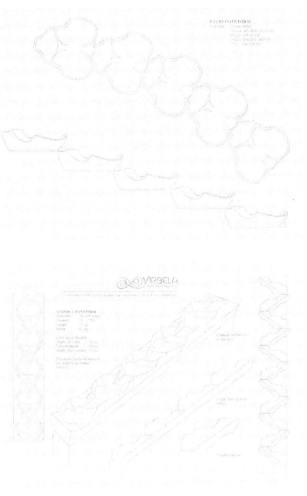

Flow rate	Optimum	Max.	Min.
lt/min	30	35	20
(gl/hr)	(396)	(462)	(264)
Gradient	68% , 34°		
Lip to lip effective length	360 mm	Overlap	158 mm
Dimensions mm	Length	Width	Height
	518	385	245
Material	Etched Granite Concrete		
Approx. Weight	25 kg		

Remarks: In plan view forms may be straight or rotated at any angle. Casts can be supplied with 15 mm dia hole for supporting pin - see drawing for location.

VIRBELA
FLOWFORMS

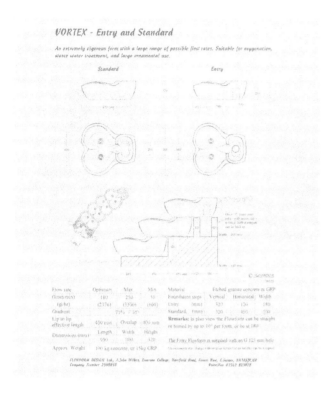

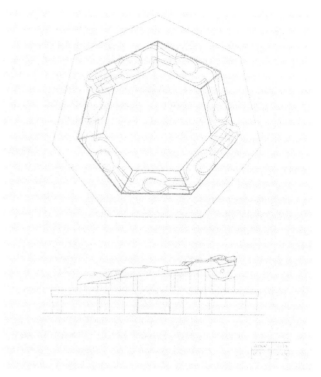

Figure 15.23 (left). Vortex
Figure 15.24 (right). Cattle trough

Glonn I*/ 93 — Wilkes /Weidmann
Designed to produce vigorous lemniscatory rhythms for mixing and water treatment purposes connected with food-processing. Capable of being stacked vertically or mounted in any curve (Fig. 15.22).

Glonn II*/ 111296 — Wilkes /Weidmann
Derived from Glonn I, a lighter version for half the flow-rate (30lt/min), can be hung on wires when made in ceramic (Fig.15.19).

Vortex*/0394 — Wilkes/Weidmann
Large volume Flowform with holes in the base for optimal vortex movement, any flow-rate up to 250lt/min+. Intended for vigorous mixing processes used for organic, biodynamic farming and sewage treatments (Fig. 15.23).

Viehtränke/ 94 — Wilkes/Weidmann*
Single cavity Flowform used on seven-sided polygon for treatment of drinking water (Fig. 15.24).

Cow-trough/95 — Wilkes/Weidmann*
Used in conjunction with drinking-water treatment cascade(s).

BD mixing (Path-curve)/93 — Wilkes/ Thomas/ Weidmann
Mathematical surfaces relating to vortex and cowhorn incorporated into the Flowform for optimizing effect in the process of mixing biodynamic preparations.

Patio II for humpmould/97 — Wilkes /Weidmann*
Simplified version of Patio I for production on hump-mould.

Helena 400 and 600 — Joiner, Iris Water & Design*
Cornelia — Joiner*
The Rose — Joiner*
Shamrock — Joiner*
*Lilla Vaaga * — Wells*
Sjoe-Liljan — Keis*
Silent Flow — Keis*
*Olympia two thirds**
LL/0797 — Schuenemann/Weidmann
LR/0797 — Schuenemann/Weidmann
SL/0797 — Weidmann

SR/0797 — Weidmann
Amended version of the original two-thirds asymmetrical Olympia design (010976), each with inlet facility and interchangeable.

Hiram/small 010697/medium 080697/large 140697 — Wilkes / Weidmann*
Large, medium and small, horizontal above and below, as random metamorphic set of Flowforms

for use in Waldorf schools. The design is intended to facilitate preparation of aggregates for stone casting, surveying, planning, foundation design and landscaping in conjunction with planting and design of steps.

Sevenfold III
Herten amended/091098 — Weidmann

*Chloen Flowform (radial)*131099 — Wilkes/Weidmann*
Small size radial Flowform with six holes in the base.

Vortex Glonn III/200801 — Wilkes/Weidmann*
Redesigning of Glonn with holes in base for 10-15lit/min.

Glonn Fulda IV/100901 — Wilkes/Weidmann*
Redesign of Glonn for 45lit/min in ceramic.

Emerson II small/010301 — Wilkes/Weidmann*

Viktors Flowform (radial 3M)/210302 — Wilkes/Weidmann* (see frontispiece vignette).
This is to be made in three radial sections and an entry unit, for flexibility of use. A third scale preliminary design to establish details will eventually be used as prototype for smaller version.

C. Range of Flowform applications

There are many purposes for which Flowforms are used. Indeed wherever water is, Flowforms can be of service, from the clearly functional or technical through to the purely aesthetic. Below is a small selection of examples from different applications. The name of the Flowform is in *italics*. 'U.K.' indicates the involvement of the Flow Design Research Group.

Biological sewage disposal, Järna Högskolan, Sweden, (pp. 97–103) U.K.

Biological sewage disposal, Hogganvik, Norway, *Akalla (Iris Water)*, (Fig. 7.21 and 7.22, p. 108f)

Children's recreation, Akalla, Sweden, *Akalla*

Children's recreation, Islington, London, U.K., *Akalla*

Municipal swimming/paddling pool features, Steinen, Germany, *Akalla (Raeck)*

School landscape features: Waldorf Schools, Düsseldorf, Germany, *Akalla (Asmussen)* (Fig. 7.20, p. 107)

Public parks, Almemeer, Amsterdam, Netherlands, *Amsterdam I & II* (Fig. 12.9, p. 156) U.K.

City centre installations, Nuneaton, U.K., *Amsterdam I Scorlewald*

Atrium features, multiple accommodation, Holland, *Ashdown II*

Patio features, Ashurst Wood, Giorgetti, U.K., *Ashdown II*

Shop interior, Reform Utrecht, Netherlands, *Ashdown II*

Small conservatory features (ceramic), Vidaraasen, Norway, *Rockery* (Fig. 15.20, p. 203)

Public sculpture features, Reutlingen, Germany, *Convex Flowform (HD)*, (Fig. 12.10, p. 157)

BD Preps mixing, Darmstadt BD Research, Germany, *Emerson*, Garden, Järna and Vortex, U.K.

Bird sanctuary, Wingshaven Sussex, U.K., *Emerson*

Clinic garden, Vidar Klinik, Sweden, *Emerson (Virbela Atelje)*

Street furniture, with road hump, Gjellerup Parken, Denmark, *Emerson (Keis)*

Private garden features, Glos. 2022, U.K., *Garden (Ebb & Flow)*

Bakery, Herrmannsdorf, Germany, *Glass Stackable, U.K.*

Water Exhibition, Graz, Stadtmuseum Water, Austria, *Glass Stackable, U.K.*

Drinking water treatment, Herrmannsdorf, Germany, *Glonn I*, U.K.

Bakery, Fulda, Germany, *Glonn II*, U.K.

Fruit Juice treatment, Olen Safteri, Norway, *Glonn II, (Fig 12.37, p. 164)* U.K.

Interior air conditioning, office areas, ING Bank, Netherlands, *Handrail* (Fig. 12.4, p. 155) U.K. & Copijn

Food processing: curing Environment, Herrmannsdorf, Germany, *Herten* U.K.

Kindergarten, Michael Hall, U.K., *Herten* (Fig 15.8, p. 196)

Private garden features, Newbury, U.K., *Herten*

School landscape features: Waldorf Schools, Hague, Netherlands, *Herten (Copijn)*

Adjustable Tunnel Flowform, Herrmannsdorf, Germany, *interchangeable blocks U.K.*

Cosmetics factory waste Body Shop, Littlehampton, U.K., *Järna (Shields)* (Fig. 7.25, p. 110)

Irrigation, Dexbach, Germany, *Järna* U.K.

Irrigation, Tenerife, Lanzerote, Spain, *Järna* U.K.

Irrigation, RDP Emerson, U.K., *Järna* with Archimedian Screw

Farm slurry treatment, Adelaide Farm, Australia, *Järna V 800 (Trousdell)*

Germination under glass, Emerson Gardens, U.K., *Lab Stackable*

Research, Inst. für Regenwurm Forsch, Austria, *Lab Stackable* U.K.

Biological sewage disposal, Järna Högskolan, Sweden, *Malmö* (Fig. 8.13–15, p. 118) U.K.

Biological sewage disposal, Kolding, Denmark, *Malmö,* (Fig 7.26, p.110) *(Keis)*

Exhibition, Waldorf Edn, Sweden, *Malmö*

Fish farming, breeding, Cackle Street, Nutley, U.K., *Malmö*

Mountain farm, Sundet, Norway, *Malmö,* (Fig 6.4, p. 89).

Biological swimming pool, Porsch Salzburg, Austria, *Malmö (HD project)*

City centre installation, Klazineveen, Netherlands, *Olympia* U.K.

Scent & Touch Garden for Blind, Kolbengraben, Ulm, Germany, *Olympia* (Fig 12.14, p. 157)

Exhibition, Mind and Body, U.K., *Olympia,* (Fig 9.5, p. 125)

Interior air conditioning, office areas, ING Amsterdam, Netherlands, *Olympia,* (Fig 12.1-4, p. 154f) U.K.

School landscape features: Waldorf Schools, Engelberg, Germany, *Olympia (HD)*

City centre installation, Hattersheim, Germany, *Porphyry (HD Project),* (Fig 12.12, p. 157).

Farm slurry treatment, Sturts Farm, Hants U.K., *Radial (Iris Water)*

City centre installation, Nuneaton U.K., *Amsterdam & Scorlewald*

Homes for the handicapped, Nutley Hall U.K., *Scorlewald* U.K.

Therapy Baths, Scorlewald Curative Home, Netherlands, *Scorlewald (Copijn)*

Homes for the handicapped, William Morris House, U.K., *Sevenfold I (Ebb & Flow)*

Biological swimming pool, Kalhamdorf, Austria, *Sevenfold II* U.K.

Christian Community, Camphill, Scotland, U.K., *Sevenfold II*

Clinic garden, Blackthorn Trust, U.K., *Sevenfold II*

Exhibition, Rieder Messe, Austria, *Sevenfold II U.K.*

Hotel features, Sheffield M1, U.K., *Sevenfold II (Ebb & Flow)*

Municipal swimming/paddling pool features, Hawkes Bay, New Zealand, *Sevenfold II (Trousdell)*

Public gardens, Chalice Well, Glastonbury, U.K., *Sevenfold II, cover*

Public gardens, Peredur Arts Centre, U.K., *Sevenfold II*

Supermarket features, shopping Mall, Lövenskog, Oslo, Norway, *Sevenfold II (Vidaraasen)*

Private garden features, Dr Douch, Sussex, U.K., *Sevenfold II 0–4*

Clinic waiting areas (cleaning of atmosphere), Skanderborg, Denmark, *Silent (Keis)* (Fig. 12.7, p. 156)

Entrance features, Directors entrance ING, Netherlands, *Single Cavity* Copijn & U.K.

Exhibition, Earls Court, U.K., *Single Cavity*

Interior air conditioning, office areas, Frankfurt Oekohaus, Germany, *Single cavity* (Fig. 12.31 & 12.34, p. 163) U.K.

Cattle drinking water, Herrmannsdorf, Germany, *Single cavity unit,* (Fig 12.38 & 12.39, p. 165) U.K.

Training College, Skanderburg, Denmark, *Skanderburg/Frankfurt (Keis)*

Farm slurry treatment, Clent pig slurry, U.K., *Slurry*

Fruit Juice treatment, Godager, Norway, *Spiral Flowform U.K.*

Foyer, Malmö office, Sweden, *Stackable (Virbela Atelje)*

Atrium features, multiple accommodation, Altersheim, Järna, Sweden, *Stensund (Virbela Atelje),* Fig 12.18 on p. 159

Clinic waiting areas (cleaning of atmosphere), Vidar Klinik, Järna, *Sweden, Stensund (Virbela Atelje)*

BD Preps mixing, Ambooti Tea, India, *Vortex (Caldes) U.K.*

BD Preps mixing, Hof Peetzig, Germany, *Vortex (Wasserwerkstatt)*

Biological sewage disposal, Herrmannsdorf, Germany, *Vortex,* Fig 7.27 & 7.28 on p. 111

Appendix 3

Scientific and technical aspects

by Nick Thomas of the Flow Design Research Group

Analysis of Flowform Parameters

Can Flowforms be designed to achieve definite rhythmic characteristics? This question becomes important when the biological effect of rhythms is known, as then particular rhythms may be more beneficial than others (see the section below on Rhythm Analysis).

In practice the proportions of the Flowform are critical in getting it to function correctly. These are gradient, dimension and shape of inlet and outlet, distance between inlet and outlet, the overall shape of the vessel, and others less easy to define. Also the natural frequency of oscillation, the operating frequency, the working volume of fluid contained and the flow rate are significant. A relationship between some of these has been found as will be described below.

In fluid dynamics the Reynolds number indicates, for a given configuration, when the transition from laminar to turbulent flow may be expected. The Flowform operates in the harmonic region between the two, and we may seek a parameter that indicates where this is. The Reynolds number is defined by:

$$R \quad = \quad \frac{rLu}{m}$$

where

r = fluid density
u = velocity of flow
m = coefficient of viscosity
L = a characteristic length

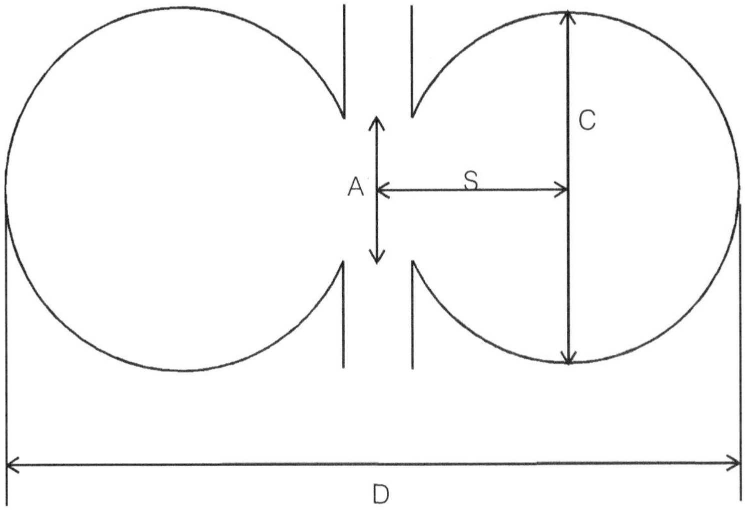

Figure 16.1

L is simply defined as the length at which turbulence sets in, for instance, for a pipe or tube, and R is typically 2000 to 4000 for pipes. Although L could be chosen as one characteristic dimension of the Flowform, such as separation of inlet and outlet, the complexity of the situation suggests a dimensionless combination of several characteristic size parameters. The following paragraphs discuss one such combination which seems significant.

The following parameters have been measured for a variety of Flowforms (Fig. 16.1):

D	Overall transverse diameter
C	Maximum chord parallel to inlet/outlet direction, averaged for asymmetrical forms
S	Separation of this chord from the centre
A	Forward aperture.

It was found that,when a rhythm is generated in a form, the function

$$L = \frac{CS}{DA}$$

tends to a constant value.

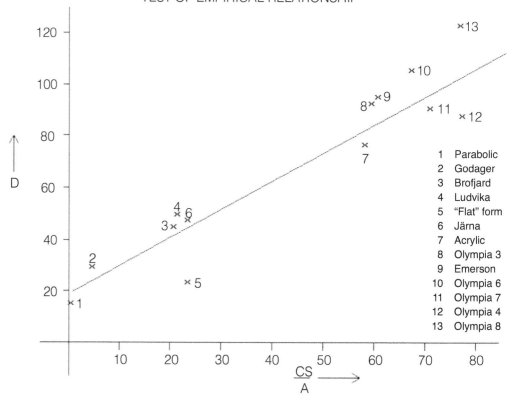

TEST OF EMPIRICAL RELATIONSHIP

1	Parabolic
2	Godager
3	Brofjard
4	Ludvika
5	"Flat" form
6	Järna
7	Acrylic
8	Olympia 3
9	Emerson
10	Olympia 6
11	Olympia 7
12	Olympia 4
13	Olympia 8

Figure 16.2　　A trendline of *CS/A* against *D* (Fig.16.2) yields a correlation coefficient of 0.96, which shows that the regression line in the graph is statistically significant. The mean value of the constant is 0.57 with a standard deviation of 0.27.

Thus for a given diameter *D* and aperture *A* we see that increase of *S* requires *C* to decrease, so that the form moves towards a more oval shape, as has arisen in practice when small forms have been designed. For larger Flowforms the ratio *C/A* is larger which means *D/S* is larger; that is, the maximum chord is proportionally closer to the centre. This suggests that the greater volume of flow in a larger form should not be turned away from the forward direction too abruptly, as would seem intuitively reasonable. It must be stressed that this is an empirical relationship derived from actual Flowforms rather than a theoretical one. It provides some insight into what proportions have been successful in practice.

In morphological terms the size of the Flowform inversely relates to the kind of Cassini curve it most resembles, small forms approaching the more oval curves and large ones the lemniscate, which suggests that a 'Cassini index' for Flowforms is definable.

The frequency of the visible pulsing is inversely proportional to the overall span D, as might be expected, with a correlation of 93%. This frequency is not exactly the same as the natural frequency of the basin; that is, if the inlet and outlet are sealed with water in the basin, and it is set in transverse oscillation, the frequency of that oscillation differs slightly from that of the pulsing of a functioning Flowform.

These findings are a start towards designing for rhythm, but are far from enough. The exact form of the inlet and outlet are very critical and no attempt has been made to analyse them. They are produced in the design process by artistic means coupled with trial and error. A question yet to be settled is how far the rhythm (see final section) is determined by the overall proportions of the form such as those explored above, and how much depends upon fine details that are difficult to quantify. A start has been made on analysing the actual forms themselves using Fourier analysis. A shape may be described in terms of harmonics, and the Fourier transform is the simplest approach to that. Legendre functions and harmonic analysis may also be applied. For example the shape of the Earth has been analysed in considerable detail by observing the motion of artificial satellites and determining shape harmonics using such transforms. Using the Fourier approach entails postulating a circular form as basic, and then departures from the circle are ascribed to 'harmonics.' The circle is at a constant distance from its centre, and variations from that may be regarded as oscillations about a mean level amenable to Fourier analysis. Then the harmonics obtained may be used to characterize the form itself, and some correlation with the dynamic rhythm sought. A typical such characteristic is shown in Figure 16.3.

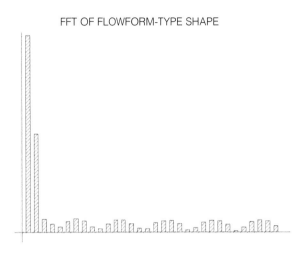

FFT OF FLOWFORM-TYPE SHAPE

Figure 16.3

Rhythm analysis

The obvious rhythmic pulsing of the Flowform suggested from the outset that any scientific significance of the method might centre on the importance of rhythm for life. Rudolf Steiner points to rhythm as a kind of bridge between the sensory and supersensory worlds. While pressure is the paradigm of physical force, and suction of the etheric, rhythm expresses the constant interweaving of the two in

Figure 16.4

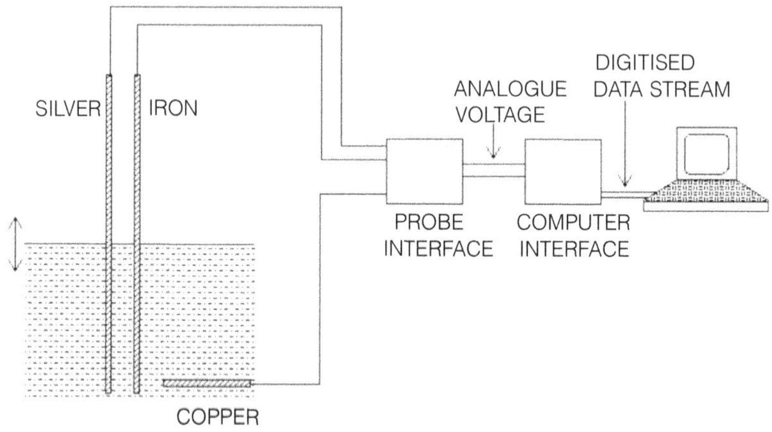

living things. It is not at the outset obvious that mere rhythm 'is good,' or 'is what is needed.' There are many qualities of rhythm, and we might expect to find some supportive of life while others may be baneful. Different Flowforms exhibit different rhythms, and it would be interesting to know their different characteristics and influences. This requires two main lines of work: to find a method to determine the power of treated water to support life, and to measure and characterize the rhythms. This section concerns the latter work. The two then need to illuminate each other.

Rhythm is not the same as mere oscillation or frequency. Rhythm in music essentially involves agogic if a mere monotonous beat is to be avoided. It is possible to analyse a rhythmic process into component simple frequencies using a mathematical procedure devised by Fourier. Thus a violin note may be exhibited on an oscilloscope, but when shorn of its harmonics or overtones a simple pitch like that of a tuning fork is obtained. The harmonics must be restored to recover the richness of the true note. They are all simple multiples of the fundamental pitch, but of varying amplitude. It is the relation between the amplitudes of the sequence of harmonics that characterizes the tone, distinguishing a violin from an oboe, for example, even though both may be playing the same note. Electronic synthesizers depend upon this principle to mimic the sounds of true instruments. It should not be imagined, however, that this analysis 'explains' the tone, a habit of thought that has crept into scientific thinking. The tone itself is borne by the physical frequencies, but is itself a definite inner experience that occurs in parallel with the physical vibration. So when Flowform rhythms are analysed we must keep in mind that the analytic method employed is intended to classify or characterize, rather than to explain.

As the water pulses in the Flowform it is possible to measure the variation in depth at a given point, say over a few minutes, and to record those measurements. Then they can be subjected to Fourier's mathematical transform to discover the component frequencies. This is simple with a digital computer, using the so-called Fast Fourier Transform procedure (or FFT for short) invented by Cooley and Tukey.

The first problem to solve was how to measure the depth conveniently, and then to convert it to a form usable by a computer. Various methods were tried after such suggestions as the use of floats had been discarded. The final method adopted relied on the potential difference that is developed between two dissimilar metals when placed in water and connected electrically. However, if an electric current is permitted to flow between them then metal will be transferred from one electrode to the other (which is the principle of electrochemical plating). This is undesirable in this case, so an electronic circuit was used to oppose the potential developed and so prevent the current from flowing. The voltage required to do that was then measured continuously (see Fig. 16.4 for a schematic diagram of the arrangement). It is interesting that a varying voltage was indeed obtained as the water level varied, and fortunately it was not realized at the time that the theory of the process did not seem to predict that! In still water no difference of voltage is obtained for different depths, but in the dynamic situation of the Flowform the voltage does indeed vary. The explanation seems to be that the Helmholtz layer is constantly disturbed, allowing the voltage to vary. Silver and iron electrodes were used, but it was found that the relation between water depth and voltage was not linear. Fig. 16.5 shows the characteristic. Addition of a third continuously immersed electrode made of copper improved the linearity sufficiently for practical work to proceed as shown in Fig. 16.6. Linearity is important or spurious frequencies appear that are nothing to do with the Flowform, but rather are due to intermodulation. These would then have to be eliminated.

Measurements of depth were controlled by an electronic interface that converted them into digital form suitable for a computer. A series of 512 precisely timed measurements was taken on each

Figure 16.5 (above).
Figure 16.6 (below).

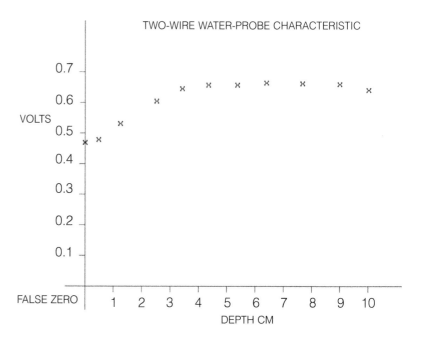

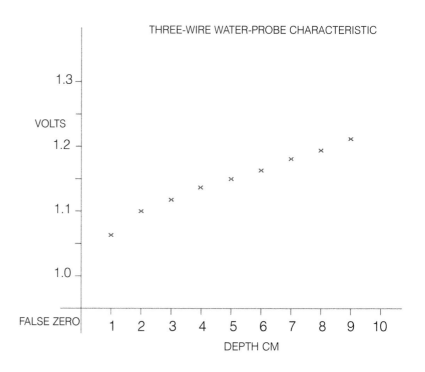

FLOWFORMS: THE RHYTHMIC POWER OF WATER

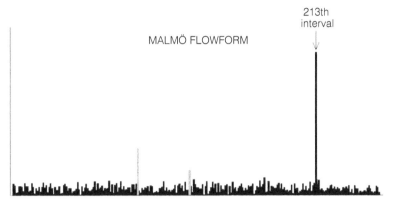

213th
interval

MALMÖ FLOWFORM

Frequency interval 0.00181686 Hz

Figure 16.7 (above).
Figure 16.8 (below).

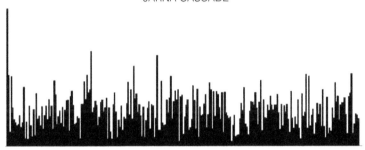

JÄRNA CASCADE

Frequency interval 0.000140892 Hz

run, and the results were stored in the computer for subsequent Fourier analysis. The interval between measurements was controllable so that the whole run could last from a few seconds to a few hours, the shorter runs being used to examine higher frequencies while the longer ones allowed lower frequencies to be checked.

Technical precautions were taken to prevent higher frequencies being 'folded' into the 'window' of interest. Initially the visible pulsing of the Flowform was timed and the result compared with the corresponding (and usually largest) frequency found by the Fourier analysis. The correspondence was excellent in all cases, which showed that the results obtained corresponded with reality.

Single Flowforms were found to have surprisingly simple rhythms, the main pulsing often comprising only one frequency. Fig. 16.7 shows an example. However, in a cascade of Flowforms the rhythms are much more complex as illustrated in Fig. 16.8. Longer term rhythms with a period of up to four minutes were noticed visually in such cases, some of which were to be seen in the Fourier analyses. This remains an open area of work as redesigned apparatus is necessary for the longer runs needed to find them, which will be much easier with modern computers. Such will be undertaken when the biological work is sufficiently advanced to enable the biological significance of the rhythms to be determined.

The analogy to musical harmonics only applies to frequencies higher than the 'fundamental' one (i.e. that corresponding to the visible pulsing). These showed the presence of surface ripples, as might be expected, but were not followed up further. Of more interest were frequencies revealed that were slower than the visible pulsing. On the first Flowform tested in this way a slower frequency was found with a ratio to the fundamental of 1 to 6 i.e. the fundamental frequency was six times as fast. Were all Flowforms the same? Would the relationship always be an integer? We did not know. Twenty Flowforms were analysed and it was found that they were all different and that integral relationships were the exception rather than the rule. Not really surprising, and also more interesting since there remained plenty of scope to compare the biological effects (if any) of different rhythms. Some might prove more beneficial than others. With such possibilities in mind work was also started on mathematically analysing the Flowform shapes themselves (see the previous section) to see how far it might eventually be possible to 'design for rhythm'. This work also awaits the further development of the biological testing without which it has little significance.

The slower frequency seems to depend upon which half of the Flowform is tested, which indicates an interaction between the form of that half and the overall pulsing. The rhythms were analysed at various different positions in the Flowform, and surprisingly different results were obtained.

The obvious place to start was on the edge of the basin where the variation was greatest , but where there was sufficient depth. However in the centre another frequency component is involved that is of equal or even greater amplitude than the obvious pulsing, an example of which is shown in Fig. 16.9.

In conclusion, this work has revealed unsuspected and interesting elements of the Flowform rhythms which may enable conclusions to be drawn as to which are most supportive of life.

For table of Flowform Characteristics and interpretation of FFT data, see page 226–27.

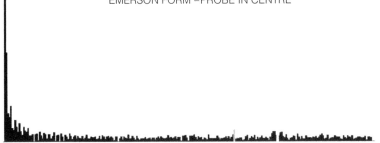

Figure 16.9.

Frequency interval 0.0008382 Hz

Path-curve surfaces and Flowforms

The idea has been present since the start of the work that the incorporation of mathematical surfaces in the design of Flowforms may be interesting. George Adams experimented with special geometrical surfaces as a way of treating water, which involved so-called path-curves. What is a path-curve? For example, it is possible to imagine the whole of space being rotated about a vertical axis. The axis will have to extend to infinity in both directions, and we may suppose that a rotation of ten degrees is made. Then every point will move along an arc of a circle, and if the transformation is repeated all points will move a further ten degrees. If this is continued a family of circles appears all centred on the axis and with their planes at right angles to it. The transformation is simple and described in terms conformable to our everyday consciousness. The circles are simple examples of path-curves, each such curve being the path followed by a point as the transformation is repeated.

In this example, although the points of space move round these circles, the circles themselves are unchanged. For a large class of transformations there exist curves (usually not circles) which are unchanged as a whole, and they are customarily called path-curves. They apply to linear transformations and were discovered by Felix Klein in the nineteenth century. George Adams studied them and found that in certain special cases they assume spiral forms either like those seen in Nature, for instance, on pine cones, or of vortical form. Lawrence Edwards, who studied projective geometry with George Adams, worked for many years to show that they describe the forms of fir cones, eggs, plant buds, water vortices — and also the human heart — with remarkable precision (Edwards 1982, 1993).

However, the real significance of these forms lies in their dual nature as mediators between space and counter-space (Thomas 1984), and for this we need to think about the dual of a curve. When we described the circular path-curves we saw that dually the planes of space moved so as to envelope cones. These cones are dual to the circles, for a circle is described by a set of points in a plane following a definite law, and a cone is enveloped by a set of planes in a point following the same law. The cone is called a 'developable', which is the formal name for a single-parameter family of planes. The name comes from the fact that a developable surface, if cut along a suitable line or curve, can be rolled out flat or 'developed' on a plane surface. This is obvious in the case of a cone. We might expect the transformation generating a path-curve also to generate a developable, and this it does. The osculating planes which touch the curve envelope a developable surface, which can also be seen as the surface made up of the tangent lines to the curve. The path-curve is then called the 'cuspidal edge' of this surface, and if the surface is cut along that curve it can be unrolled and laid out flat. Thus the path-curve enjoys an intimate relation between the point based view and the dual planar approach, and George Adams suggested that for this reason it has significance as a mediator between space and counter space. But it is the dynamic character of the curve that is intended — the curve arising from a process — that is important, not merely its spatial form. If indeed living organisms incorporate etheric forces, and these curves are seen so often in Nature, then it seems that they play a fundamental role in the way the living comes to manifestation in space. Lawrence Edwards has demonstrated this in his practical research. The key is to grasp that an interweaving of two different kinds of space is involved here, which is why both the point and plane aspects need be understood and related.

A path-curve for an axially symmetrical surface is mathematically described by two parameters (Edwards 1982): λ (lambda) which determines in the case of eggs how 'pointed' they are (Fig. 16.10), and ϵ (epsilon) which determines how steeply they spiral. λ determines the shape in any plane through the axis; that is, of the vertical cross sections such as those in the figure. However, it is also possible to have path-curve surfaces with horizontal cross sections

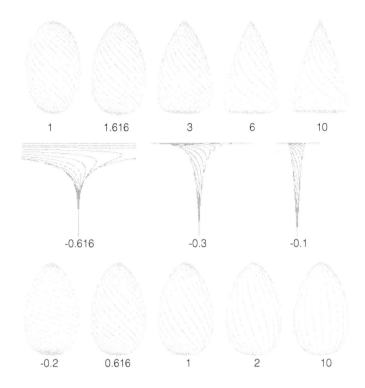

which are spirals instead of circles, in which case the path-curves require a third parameter. This is μ (mu), which is the λ-value of the vertical cross sections (Edwards 1982), while λ proper is now the parameter of the curve obtained if the path-curve cuts a vertical plane and is rotated about the axis so that the point in which it cuts that plane moves along either an egg or vortex profile. Although this sounds complicated, and is difficult to draw, it is most readily grasped in thought if a horizontal logarithmic spiral is visualized which moves upwards and rotates at the same time, so that it forms a surface (Fig. 16.11). If we start with such a spiral and a vertical path-curve which goes through it, then as it moves upwards — remaining horizontal — it rotates so as always to intersect that path-curve.

Now life and rhythm are intimately related, and the interworking of space and counter-space seems to be itself rhythmic as we transform back and forth between them, and so is the functioning of the Flowform. Hence the motive to see if water treated in this way is better able to support life. In view of the relation of path-curves to this, it is expected that incorporation of path-curve surfaces into

Figure 16.10 (left).
Figure 16.11 (below).

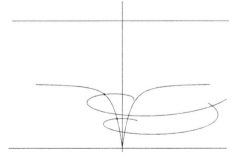

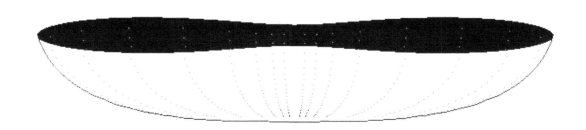

Figure 16.12. Flowforms may enhance their effectiveness. Now what do we mean by a path-curve surface? A transformation yielding egg path-curves does not especially produce surfaces, but if another path-curve such as a circle is placed in space which is related to the same reference system then the path-curves which pass through it will form a surface. An egg surface arises when a circle centred on the axis of symmetry is used in this way, which is typical of bird's eggs and plant buds. We may also use other curves, and Fig. 16.12 shows that Flowform-like surfaces may be obtained when path-curves are passed through Cassini curves. The Cassini curve tends to arise in more complicated transformations called pivot transforms (Edwards 1993). These comprise a more intimate relationship between point-like and plane-like forms, especially where seed pods arise. A Flowform could also arise in this way.

We see that forms may be related to processes, and hence to the cosmic life forces upon which all living things depend. This is particularly well illustrated in the case of the water vortex which only exists as a process or movement of water, and which is also a path-curve (Edwards 1993). The question to be explored further here concerns which forms allow water to flow in such a way that it is brought into relationship with the cosmos. This has particular relevance to biodynamic agriculture which seeks such relationships.

A surface like a vortex has negative curvature, which means that at each point on its surface there are two tangents along which it has zero curvature. The ruled hyperboloid is a simple example (Fig 16.13), for two straight lines in the surface pass though every point. The curvature changes as the tangent lines at the point pass through one of these special tangents, in the sense that it switches from being directed inwards to outwards. The directions of the special tangents are called asymptotic directions. This is also true of a vortex, but in this case there are no straight lines in the surface, so if we follow the asymptotic direction from point to point we obtain a curve instead of a straight line, and such curves are called asymptotic curves. These curves are themselves path-curves in the case of path-curve surfaces. George Adams investigated them, and special surfaces were constructed for experimental purposes.

A project was carried out to construct a surface, suitable for incorporating in a Flowform, that is woven of two special sets of asymptotic curves. Since a spiral path-curve surface has two asymptotic directions at every point, its surface contains two sets of asymptotic curves spiraling round it in opposite directions. If the surface is axially symmetric like a vortex then the two sets are essentially the same apart from the direction of spiralling, but for more complicated surfaces with spiral instead of circular horizontal cross-sections the two sets have different parameters (Edwards 1982). It can be shown (Thomas 1991) that if the two λs of these asymptotic curves are given and fulfil certain conditions then the resulting surface is unique. The application to biodynamic mixers suggests that the λ-values for the water vortex and the cow horn be used, and this has been tried. Segments of the resulting spiral surface can be brought together to give the fundamental Flowform shape, but the details of inlet and outlet must be found empirically. (see Figs 12.40 & 12.41 on p. 165).

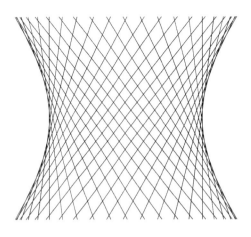

Figure 16.13.

Appendix 4

The Virbela Rhythm Research Institute

The Flow Design Research Group was founded in 1975 by Nigel Wells, Nick Thomas and John Wilkes, its essential aim being to pursue investigation of rhythmic processes generated in water through the use of the Flowform method. For many years we continued our work with Flowforms in a temporary building at Emerson College, Sussex, while we attempted to build up resources for our building endeavour which had ceased through lack of funds. In recent times major donations made it possible to reactivate the building process, and the new Institute building was brought to its final stages in 2002 within an overall plan of College expansion. The institute maintains independent scientific and financial management within the Emerson College Charitable Trust.

The new building offers about 270 square metres of working space on three floors. The construction is as environmentally friendly as we could make it. The ground floor consists of the main workshop studio with accompanying spaces for demonstrations and test methods that need a hard stable surface. Adjacent is an area under glass at the front of the building which can serve a number of uses. We also have a ceramic studio with kiln. The first floor has the main seminar space with two further studios and a small office. The upper floor provides us with important extra space for design and other activities.

We intend eventually to collect as much water on site from the roof and surrounding slopes, a reservoir having been built for this purpose from the original excavation material. In conjunction with this, Flowform demonstrations will be installed as an integrated landscape around the building, much of this work being done within the framework of training programmes.

Flowform Design Research

Since the discovery of the Flowform method in 1970, about one hundred Flowform designs of different types have been made and used. There are some fifty types (see Appendix 2) of which only a number have been investigated. Wherever possible, projects are utilized to create designs with new capacities and uses.

We remain interested in using a whole range of materials which are continually under consideration. Reconstituted stone has been the normal material for reproduction for obvious economic reasons. Glass-reinforced cement can be very useful but is rather expensive. For transport it may be convenient, but on site such products have to be bolted down and are vulnerable to breakage unless in a protected environment. Glass reinforced resin is also very useful so long as chemical effects are neutralized or at least minimized by heat treatment. Glass is a material we have used for pressings very successfully but is not easy to have reproduced due to cost. Other methods with glass are being investigated. Metal for casting is good but rarely commissioned; however designs exist and we are looking forward to carrying them out when opportunity arises. Other casting materials such as refractory cements are of interest and await further investigation.

Ceramic materials are being used and provide a good neutral material for situations where water quality is of paramount importance such as in drinking water and food processing. A really interesting area is that of natural stone which can enhance the quality of water and its mineral content. In all these cases, the positive effect of materials for given purposes is important.

Our future work will be largely concerned with the optimization of effects upon the water: firstly through the effect of rhythms and their different frequencies, but then also from the intimate quality of the designed surfaces whether empirically or mathematically founded; then the positive influence of materials; and finally the times and duration of treatments. As explained in the text, water is a medium which mediates influences of the environment to the organism, and this is a vast area awaiting further investigation. The latest realm to be investigated concerns the additional influence of architectural or geometrical spaces in which rhythmical cascading process are housed (see page 167).

Scientific Research

As already described, we are essentially involved with the investigation of rhythmic processes generated in water through the use of the Flowform method and other closely related technologies. Scientific work is largely connected with establishing the nature of effects achieved through treatments with the help of a range of methods. This research indicates qualitative effects relating to an improvement in water's life-sustaining capacities. It is by means of very specific proportions, that water streaming through the vessels is resisted and thus brought into pulsing vortical and lemniscatory rhythms. This principle of resistance is active wherever rhythms are generated. There seems to be something very special at work here. Over the years one realizes that there must be profound forces active in the establishment of the correct proportions. For me it has always been a question of creating the right situations in which processes can develop in nature's own terms. The more we can demonstrate, observe and understand water's intimate movements, the more we will understand the subtleties involved.

Conferences and Workshops

There have been many requests over the years to offer courses of different kinds. Up to now, these have been held in other locations internationally. Here in Sussex this will now be increasingly possible with permanent spaces becoming available. Given our new facilities, we shall be setting up more sophisticated demonstration and experimental work with water phenomenon. This apparatus will not only be there to show participants what goes on within a body of water, but also for research purposes.

There exist various target groups for gatherings: people simply interested in water themes; those interested in medicine and in the healing properties of water; food processing organizations which are intent on enhancing the nutrient value of their products; those interested in working with biological systems to purify and revitalize water after it has been used for industrial and domestic purposes; farmers and gardeners who are using water to support the life of plants and animals; teachers who are educating children and need help and stimulation to enter the whole mysterious world of water to create a new consciousness for coming generations. Water in science and art could become an important field within new curricula. These are just a few of the many areas needing support and perhaps through these activities we can generate new ideas together.

Virbela International Association

Over the years a wide range of contacts and relationships has grown up across some thirty countries. This has given rise to a network of international associates working in close collaboration and under the auspices of the Institute. Two sections of the Association are envisaged as it undergoes improved organization and communication: those practically involved and those within a wider circle who express the wish to support the work of the Institute.

The Flow Design Research Group will constitute a focus point for the Association. At present the members most closely connected with this Group are Nick Thomas, Nick Weidmann, Costantino Giorgetti, Judyth Sassoon, Tadeu Caldes, Thomas Wilkes, Thomas Hoffmann, Jochen Schwuchow and John Wilkes. Others may well join later.

As already mentioned, a wider growup of collaborators around the world who are concerned with the use and distribution of Flowforms are looked upon as Associates. A few are also accredited as Flowform designers: Nigel Wells (Sweden), Iain Trousdell (New Zealand), Mark Baxter (Australia), Andrew Joiner (U.K.), Herbert Dreiseitl (Germany), Michael Monzies (France), Hanne Keis (Denmark), Nick Weidmann (U.K.), Christopher Hecht (U.S.A.), Thomas Hoffmann (Germany).

On July 6–7, 2002, an international meeting of Associates was called, mainly initiated by Simon Charter of Ruskin Mill (Ebb and Flow), in conjunction with the Water Conference being held at Emerson College between July 7-13. Part of the Associates' meeting was able to assemble in the unfinished Institute building and was thus the first event to be held there. Through this July 7 meeting it was possible to establish the beginning of an Associates Institute Support Group which takes on some financial responsibility for the maintenance of the Institute. The intention is to invite all Associates to join this Support Group in the hope that all will be part of a more intensive exchange of correspondence.

I consider it important to mention those present on July 7 at this historical founding moment, who were: Nigel Wells (Sweden), Hanne Keis, Juergen Keis (Denmark), Michael Monzies (France), Juergen Uhlvund (Norway), Christopher Hecht (U.S.A.), Simon Charter (U.K.), Peter Müller (Germany), John Wilkes (U.K.). Costantino Giorgetti, Aonghus Gordon and Nick Weidmann were unable to attend fully, and the following sent their apologies: Judyth Sassoon, Mark Moodie, Jane Shields, David Shields, Andrew Joiner and Philip Kilner (U.K.), Will Browne (Norway), Iain Trousdell (N.Z.), Christopher Mann (U.S.A.), Sven Schuenemann (U.S.A.), Mark Baxter (Australia).

A still wider group of supporters is envisaged, able to offer advice and support of different kinds as well as financial help through a system of contributions, to be known as Friends of the Virbela Rhythm Research Institute, Emerson College Trust. Members will receive regular email Newsletters from the Institute. Those who are interested in becoming members of this group should contact the following website for further details:
www.emerson.org.uk
or see
www.healingwaterinstitute.org

Interpretation of FFT Data

In each case the figure (Fig. 16.7–16.9) has as its ordinate the power amplitude of the frequency components, and as abscissa the frequency. The left hand end is zero Hz (i.e. cycles per second) and the right hand end is 256 multiplied by the frequency interval printed at the top (which was calculated by the computer). Intermediate frequencies are calculated proportionally.

For example the obvious pulsing of the Flowform in the case of Fig. 16.7 was measured with a stop watch and found to be 2.6 seconds between maxima, or 0.38 Hz. The greatest amplitude frequency bar in the diagram has a frequency of 213 multiplied by 0.00181686 Hz which equals 0.387 Hz.

In each case a frequency filter was used to prevent aliasing, set to at least twice the highest frequency on the graph (Nyquist's principle) e.g. to 0.8 Hz for Fig. 16.7. Some spread of the fundamental is evident in most cases as it was not feasible with the experimental arrangement used to ensure that the length of the run precisely equalled a whole number of cycles of the fundamental frequency. This would be necessary for the mathematical transform to obtain a single 'spike' for that frequency, otherwise it 'smears' it.

Table A1: Flowform Characteristics

Flowform	Position	f_0 Hz	f_1 Hz	f_2 Hz	f_0/f_1	Amplitude	Diameter
Lab. form	4	1.74	0.29		6.0	1.48	14.5
Malmö	3	0.36	0.023		15.65	0.73	94
Malmö	1	0.381	0.252		1.51	7.3	94
Malmö	2	0.382	0.068	11.66	5.62		
Brofjord 1	4	0.775	0.113	2.314	6.86	7.14	43.8
Acrylic	4	0.429	0.135	14.653	3.18	1.05	77
Cogut RH	4	0.478	0.266	12.464	1.80	2.9	67
Cogut LH	2	0.476	0.289	7.042	1.65	3.85	67
Akalla Large	1	0.221	0.046	11.985	4.80	6.67	158
Akalla Left	5	0.22	0.055	0	4.00	0.78	158
Akalla Medium	4	0.558	0.311	7.782	1.79	4.76	67
Godager	4	0.903	0.27	3.84	3.34	3.57	29
Olympia 7 RH	4	0.435	0.124	12.735	3.51	4.55	92
Olympia 7 LH	6	0.436				9.266	92
Olympia 6 RH	7	0.325				12.218	106
Olympia 6 LH	8	0.351	0.033	1.144	10.64	2.94	106
Järna single	4	0.668	0.249	13.54	2.68	6.45	47.5
repeat	4	0.672	0.304		2.21	6.25	47.5
Järna cascade 9th	4	0.686	0.0156		44.00	1.41	47.5
repeat	4	0.695	0.0224		31.03	0.41	47.5
Olympia 2	4	0.218	0.045	7.262	4.84	5.41	146
Olympia 3	4	0.363	0.099	13.277	3.67	2.67	93
Olympia 4	1	0.361	0.097	6.974	3.72	1.0	87
'Flat form'	4	0.028	0.453	8.48	2.27	1.85	24.3
Cylindrical 1	4	0.84	0.0196	8.24	42.86	0.78	37.5
Cylindrical 2	4	0.569	0.203	6.385	2.80	2.18	72

f_0 = Fundamental frequency (observable pulsing)
f_1 = Slower secondary frequency
f_2 = Higher secondary frequency
f_0/f_1 = f_0 divided by f_1
Dia = Total span orthogonal to main flow direction

Endnotes

1 As Rudolf Steiner points out, we can discern in the form and development of each plant, the purest presentation of an organism:

> What is alive is a self-contained wholeness which brings forth its own states. The parts when put next to each other, or the transient physical stages of a living entity reveal an interrelationship which does not appear to be caused by the sense-perceptible properties of those parts. Nor is it caused by a mechanical-causal dependence upon what has appeared earlier and what follows later, but rather by a controlling higher principle, which lies beyond the parts and stages. In a living organism there is development of one phase out of another, a changing over from one stage into another, there is no finished existence of the detail, but rather a continual becoming. (Rudolf Steiner. *Introduction to Goethe's Scientific Writings*)

2 There are far more sophisticated methods of carrying out this experiment, the results of which are illustrated in Theodor Schwenk's book *Sensitive Chaos*.

3. To give further experimental details, a shallow tank is needed, the size of which can be changed at will to vary the depth of fluid. It is constructed of four pieces of wood with a section something like 3 x 4 cm, two short say 30 cm and two long at 80 cm. These can be placed on a flat board of adequate size, at 90° to each other overlapping alternately. A piece of black plastic sheet is placed over the whole and pressed into the inner cavity. Into this a mixture of water and glycerine (about 1 to 3) is poured. Viscosity can also be increased (more cheaply) with the use of a sugar syrup. On to this surface the lycopodium powder is dusted. This can be done by using a 35 mm film cassette with a piece of muslin or gauze stretched over the top and secured. This is turned upside down and tapped on the bottom while moving over the water surface. It can be applied in a straight line down the middle, two lines, one either side, or spread evenly. A straight-line movement is then made with an object down the centre of the tank.

It is also possible to use an oil-based, thinned colour on the surface (as for marbling) which allows the results to be lifted off with the careful laying on of a piece of paper.

4 See J. Bockemühl, 'Bildebewegungen im Laubblatberich höherer Pflanzen,' *Elemente der Naturwissenschaft*, no.4, 1966, pp.7–23.

5. Importantly, I consider this sensitive harmonic realm to be that within which all of nature's forms appear. Every form in nature represents a single balanced condition, either substantially harder or softer, between contractive and expansive forces.

6. The Institut für Strömungswissenschaften (Flow Research Institute) was set up in 1961 in Herrischried by a team of scientists concerned at that time with mixing processes involved in the production of a cancer remedy: Dr Leroi who later founded the Lukas Klinik in Arlesheim, Switzerland, wanted to know more about the movement of liquids; Theodor Schwenk, a hydrodynamics engineer and director of the laboratory for the pharmaceutical company Weleda, with his assistant Helga Brasch (later Schwenk); Dr Georg Unger, mathematican and Director of the Mathematical Astronomical Section at the Goetheanum; George Adams, whose path curve re-

search provided the important stimulus for this initiative, and his assistant Olive Whicher, both of the Goethean Science Foundation, Clent, U.K.. The Institute was funded by the industrialist Dr Hans Voith of Heidenheim and his daughter Martina, both essentially involved in founding and supporting activity and, together with Herr von Zabarn, members of the umbrella Association for Movement Research (Verein für Bewegungsforschung).

7. The vertebral column consists of seven groups of bones: cranial, atlas/axis, cervical, dorsal, lumbar, sacral, caudal. The bones within each group are closely related in function.

8. Very early on, I became interested in the possibility of cavities made of a flexible skin, for instance a silicon rubber that could be suspended in a fluid medium. The proportions would be such as to bring water flowing through the system into a pulsing movement. The effect would demonstrate a pulsing movement through a kind of organ, generated purely by the flow. We could compare this effect with the fact that blood passed through a heart which has been separated from the organism for even as long as a day still generates a pulse (see p.85). The proportions would be difficult to establish because each attempt would have to be completely executed before the function could be tested out. Such an approach remains to be achieved, and might well still be attempted.

An enclosed inflexible form could also be attempted, preferably in glass or other transparent material. The water surface would have to remain free to move so that the oscillation would carry it up and over to the central stream into which it would again fall. There would be many interesting questions to investigate, for instance concerning the movement of the air in the cavities. The clear plastic Stackable Flowform is a first step in this direction.

9. When an object or method has an existence it needs a name. It was nevertheless some years before this really became a problem that had to be solved for the Flow Research. The first name that came into use in 1973 was 'Vortex-eight' which literally described the phenomenon. This was not entirely satisfactory; but it seemed impossible to find an apt solution (I found out only later that the most august groups of scholars are commissioned to devise names for products!)

A humorous intermediate 'Vortaflow' was used for a leaflet and after much searching, in 1975, I made up another name, 'Virbela' (the emphasis on the first syllable).

Only later still did I start using the word 'Flowform' for actual vessels. This is an easy descriptive word which can be used internationally. Unfortunately it is very often written in every imaginable way but the right one!

Once I had decided upon a name, the next problem was to find a distinctive way of writing it. It so happened that Christopher Mann was in close contact with Walter Roggenkamp and between them they came up with a more professional presentation of my original design which I had begun to use already as a letterhead with logo (see Fig. 5.12 on p. 69).

10. During 2004 in Giubiasco, Switzerland, ten Vortex Flowforms have been installed to treat spring water destined for municipal supplies. In order to extract energy from the water after falling hundreds of metres it is passed through hydro-electric turbines. Due to this it becomes loaded with electomagnetic frequencies making it aggressive, even affecting the concrete reservoir. Tests confirmed a neutralisation of the condition after rhythmical vortical treatment. See an article by Peter Gross referring to Prof. W.Ludwig *www.comeweb.de* (Oct/04). Quote : "In fact the simplest and at the same time a Method without side effects is an intensively repeated treatment through alternating vortices." This is precisely what we can achieve with Flowforms.

11. My main assistant with casting during this summer was Reikart Thiesson, from the Netherlands.

12. Attempts have continually been made

to assemble data on the functioning of these sewage systems. Data has been accumulated and a report was prepared by Christian Schönberger and Professor Christian Liess, commissioned by Atelier Dreiseitl in conjunction with the international working-group for water questions, Arbeitskreis für Strömungsforschung.

Reports were prepared by Prof. Petter Jenssen and colleagues on the Solborg system. See Jenssen, Krogstad and Maehlum, Report on Solborg, Ecological Engineering for Waste Water Treatment Conference 1991. See also Uwe Burka and Peter Lawrence, 'A new community approach to waste-water treatment with higher water plants,' report on the Oaklands Park Purification System (occasional paper, no date).

13. Research work carried out by the author with Nigel Wells, Paul van Dÿk, Patrick Stolfo, Philip Marchand and others. Small scale models were made in an attempt to solve many of the questions posed.

14. Support for the mould-making was guaranteed by Chris Hall, the exhibition designer, which was very much appreciated.

Nick Weidmann joined the work at this time, to make the end reservoir form and for the final preparation of the prototypes. He went over to Jan Grégoire in France in whose studio the nine moulds were being made.

15. See for instance the response to the work of Dr Jacques Benvéniste, described in the book *Memory of Water* by Michel Schiff.

16. For a fuller account of Goethe's approach to scientific observation, see *The Wholeness of Nature*, by Henri Bortoft.

17. In subsequent research, the Flowform effects were to be compared with the relatively chaotic movements produced in a channel bed filled with gravel, stones and small boulders. This would have involved a conversion of the step cascade channel. This would have allowed a study of the life-supporting capacity of the water through the use of specific rhythms compared with the natural regenerative processes of a stream bed. This work has not yet been carried out.

18. For fuller accounts of these methods, see the following: *Bewegungsformen des Wassers*, Theodor Schwenk; and more recently *Sensibles Wasser No. 6*. Wolfran Schwenk Herrischried 2001; *Die Kupferchlorid-Kristallisation*, A & O Selawry; *Agriculture of Tomorrow*, Eugen and Lily Kolisko.

19. See Earth Policy Institute article of November 22, 2001, on website *http:\\earth-policy.org*

20. These two strands of work initiated by Adams and Schwenk were subsequently amalgamated in the Institut für Strömungswissenschaften at Herrischried in 1961 (see note 6 above). It was during this period that the first path-curve surfaces were prepared for investigation with water. Many basic models are still available for study. A report on this work is available from Peter Nantke. There is one piece of apparatus that I conceived and made at that time, in the form of a ceramic spiralling tube based on a vortex surface, through which water can be drawn up by means of rotation, the possible effect of which still needs investigation. A larger available edition created during the 1980s has still to be set up in a fish-breeding system.

Sources and Further Reading

*books in which Flowforms appear

Adams Kaufmann, George. 1934. *Strahlende Weltgestaltung.* Dornach. Mathematisch-Astronomische Sektion.

Adams, George. 1979. *The Lemniscatory Ruled Surface in Space and Counterspace.* London. Rudolf Steiner Press.

Adams, George and Whicher, Olive. 1980. *The Plant between Sun and Earth.* London. Rudolf Steiner Press.

Agematsu, Yuji. 1998. *Steiner Architecture.* Tokyo. Codex. *

Alexanderson, Olaf. 1982. *Living Water.* Bath. Gateway Books.*

Allen, Joan. 1990. *Living Buildings.* Aberdeen. Camphill Architects.*

Ash, David and Hewitt, Peter. 1990. *Science of the Gods.* Bath. Gateway Books.

—. 1994. *The Vortex, Key to Future Science.* Bath. Gateway.

Biodynamics. 1989. Auckland. Random House & New Zealand Biodynamic Ass.*

Bockemühl, J. 1967. Das Ganze im Teil [The whole in the parts]. *Elemente der Naturwissenschaften* 6. 1–8.

Bortoft, Henri. 1996. *The Wholeness of Nature.* New York. Lindisfarne Books & Edinburgh. Floris Books.

Browne, W. and Jenssen, Petter D. 2001 Nov. Exceeding tertiary standards with a pond/reedbed system in Norway. Paper given at Ecological Engineering for Landscape Services and Products conference at Lincoln University. New Zealand.

Bunyard, Peter. 1977 Oct. Imitating nature to treat sewage. *New Scientist.*

—. 1978 Jan. *The Ecologist.*

—. Forum for a better world. PHP Tokyo 9.4.

Clover, Charles and HRH Prince Charles. 1997. *Highgrove — Portrait of an Estate.* London. Weidenfeld and Nicholson.*

Coates, Callum. 1996. Living Energies. Bath. Gateway.

—. 1998a. *Nature as Teacher.* Bath. Gateway

—. 1998b. *The Water Wizard.* Bath. Gateway

—. 2000a. *The Fertile Earth.* Bath. Gateway.

—. 2000b. *The Energy Evolution.* Bath. Gateway

Coates, Gary. 1997. *Erik Asmussen Architect.* Stockholm. Byggför-laget.*

Davis, Joan. 1995. *Ist Wasser mehr als H₂O?* Lucerne. Erni-Stiftung.

Day, Christopher. 1990. *Places of the Soul.* London. Aquarian.*

De Jonge, Gerdien B. 1982. Orienterend onderzoek naar de invloed van stromingsbewegingen in zgn. Wirbela Flowforms op het zelf-reinigend vermogen van organisch belast slootwater. *Ondersoeksverslag 1978–81* [Research influence of Flowforms on treatment of sewage]. Wirbela Waterprojekt, Warmonderhof, Kerk-Avezaath.

Dejonghe, Walter. 1991. *Water Wijzer.* Baarn. Bigot.*

Dienes, Gerhard M. and Leitgeb, Franz. 1990 *Wasser.* Graz. Leykam.*

Dreiseitl, H.; Grau, D. and Ludwig, K. 2001. *Waterscapes.* Basle. Birkhauser.*

Edmunds, L Francis. 1997. *Quest for Meaning.* New York. Continuum.

Edwards, Lawrence. 1982. *The Field of Form.* Edinburgh. Floris Books.

—. 1993. *The Vortex of Life.* Edinburgh. Floris Books.

Eriksson, Nils-Erik. 1978. *Fisken* [Fishing]. Stockholm. Eriksson, Johnson och LT's Forlag.

Fant, Åke; Klingborg, Arne, and Wilkes, John. 1969. *Holzplastik Rudolf Steiners.* Dornach. Philosophisch Anthroposophischer Verlag.

—. 1975. *Rudolf Steiner's Sculpture.* London. Rudolf Steiner Press.

Geiger, W. and Dreiseitl, H. 1995. *Neue Wege für das Regenwasser.* Munich. Oldenbourg.

Goethe, J.W. von. 1960. *Metamorphose der Pflanze.* Stuttgart. Freies Geistesleben.

Grant, Nick; Moodie, Mark and Weedon, Chris. 1996. *Sewage Solutions.* Machynlleth. Centre for Alternative Technology.*

Griggs, Barbara. 2001. *Reinventing Eden.* London. Quadrille.*

Grohmann, Gerbert. 1990. *Metamorphosen im Planzenreich* [Metamorphoses in the plant kingdom]. Stuttgart. Freies Geistesleben.

Hall, Alan. 1997. *Water, Electricity and Health.* Stroud. Hawthorn Press.

Hancock, Graham. 1995. *Fingerprints of the Gods.* London. Mandarin.

Helliwell, Tanis. 1997. *Summer with the Leprechauns.* California. Blue Dolphin Publishing Inc.

Hoesch, Alexandra *et al.* 1992. *Lebendiges Wasser* [Living water]. Munich. Schweisfurth-Stiftung.

Honauer, Urs. 1998. *Wasser, die geheimnisvolle Energie.* Munich. Hugendubel.*

Julius, Frits. 1969. *Metamorphose: Ein Schlüssel zum Verständnis von Pflanzenwuchs und Menschenleben.* Stuttgart. Mellinger.

Klingborg, Arne. 1998. *En Trägardspark i Södermanland.* Stockholm. Wahlström & Widstrand.*

Kronberger, Hans, and Lattacher, Siegbert. 1995. *On the Track of Water's Secret.* Arizona. Wishland.

Lattacher, Siegbert. 1999. *Viktor Schauberger.* Steyr. Ennsthaler.

Lebendige Erde. 1996 May. Schwerpunkt Wasser.

Leopold, Luna B. and Davis, Kenneth S. 1971. *Water.* Amsterdam. Time-Life.

Locher-Ernst, Louis. 1957. *Raum und Gegenraum.* Dornach. Philosophisch Anthroposophischer Verlag.

Marinelli, R. 1996. The Heart is not a Pump. *Frontier Perspectives.* 5.1.

—. 1998. Das Herz ist keine Pumpe. *Raum und Zeit.* No. 93.

Mæhlum, Trond. 1991. Økologisk avløpsrensing [ecological drainage]. (1. Bruk av konstruerte våtmarker for rensing av avløpsvann i Norge; 2. Effekten av strømningsformer (Flowforms) for lufting biodammer om vinteren; 3. Wastewater treatment by constructed wetlands in Norwegian climate: Pretreatment and optimal design.) Jordforsk, Norges Landbrukshøgskole.*

Mendelsohn, Martin. 1928. *Das Herz — ein sekundäres Organ* [The heart, a secondary organ]. Berlin. Axel Juncker Verlag.

Naydler, Jeremy (ed.) 1996. *Goethe on Science.* Edinburgh. Floris Books.

Nelson, Carolyn T. 1987. Wheat growth in light experiments, 1986. *Flow Design Research Group Internal Report.* Nos 1 & 2. Emerson College. Forest Row.

Pearson, David. 1989. *The Natural House Book.* London. Conrad Octopus.*

—. 1994. *Earth to Spirit.* Stroud. Gaia Books.*

—. 2001. *New Organic Architecture. The Breaking Wave.* Stroud. Gaia Books.*

Peter, Heinz-Michael. 1994. Self-regeneration of the river Mettma. *Sensibles Wasser.* No. 4.

Pogacnik, Marko. *Elementarwesen.* Munich. Knaur.

Proctor, Peter. 1997. *Grasp the Nettle.* Auckland. Random House.*

Radlberger, Claus. 1999. *Der Hyperbolische Kegel.* Bad Ischl. PKS.

Raum & Zeit. 1992. Special Issue No. 6. Electro-Smog.

—. 1994. Special Issue No. 7. Freie Energie.

Ryrie, Charlie. 1999. *The Healing Energies of Water.* Stroud. Gaia Books.*

Schatz, Paul. 1975. *Rhythmusforschung und Technik.* Stuttgart. Freies Geistesleben.

Schauberger, Victor, *see* Coats, Callum.

Schiff, Michel. 1995. *The Memory of Water: Homeopathy and the Battle of Ideas in the New Science.* London. Thorsons.

Schikorr, Freya. 1990. Wheat germination responses to Flowform treated water. *Star and Furrow.* 74.15–22.

Schneider, Peter. 1973. Ab-Fluss oder Ab-Wasser, ein inner-Welt oder Um-Weltproblem? [Sewage and environment]. *Elemente der Naturwissenschaft.* 19.25–36.

Schmidt, Karl. 1892. Über Herzstoss und Pulskurven [On heart-beat and pulse curves]. *Wiener Mediz. Wochenschrift.* 15.

Schönberger, Christian and Liess, Christian. 1995. *Wirksamkeit der Flowforms – Zusammenstellung und Auswertung der bis 1994 durchgeführten Untersuchungen über die Wirkung der Virbela-Flowforms* [Evaluation of efficiency of Flowforms]. Überlingen. Atelier Dreiseitl.

Schwenk, Theodor. 1995. *Sensitive Chaos: the Creation of Flowing Forms in Water and Air.* London. Rudolf Steiner Press

—. 1967. *Bewegungsformen des Wassers.* Stuttgart. Freies Geistesleben.

Schwenk, Theodor and Schwenk, Wolfram. 1989. *Water: the Element of Life.* New York. Anthroposophic Press.

Seamon, David and Zajonc, Arthur. 1998. (eds.) *Goethe's Way of Science: a Phenomenology of Nature.* State University of New York Press.*

Sheldrake, Rupert. 1981. A *New Science of Life.* London. Blond & Briggs.

Smith, Cyril. 1989. *The Electromagnetic Man.* London. Dent & Sons.

Sokolina, Anna. 2001. *Architektura i Antroposofia.* [Architecture and Anthroposophy]. Moscow. KMK Scientific Press Ltd. *

Stauffer, Julie 1996. *Safe to Drink?* Machynlleth. Centre for Alternative Technology.

Steiner, Rudolf. 1972. *Man: Hieroglyph of the Universe.* London.

Rudolf Steiner Press. [Translated from lectures April/May 1920. GA 201].

—. 1975. Spiritual Science and Medicine. London. Rudolf Steiner Press. [Translated from lectures March/April 1920. GA 312].

—. 1980. *Second Scientific Course*: *Warmth course*. New York. Mercury Press. [Translated from lectures March 1920. GA 321].

—. 1990. *The Riddle of Humanity*. London. Rudolf Steiner Press. [Translated from lectures 1916. GA 170].

—. 1997. *An Outline of Esoteric Science*. New York. Anthroposophic Press. [GA 13].

Strid, Martin. 1984. *Rhythmik Strömning: en Studie av Virbela flödesformer* [Rhythmic flow]. Luleå. Tekniska Högskolan.

Strube, Jurgen and Stolz, Peter. 1999. Verbesserte Wasserqualität zum Brotbacken durch simulierten Bergbach mit Flowformen, [Improved quality of water for baking bread] Dipperz. Kwalis Qualitätsforschung Fulda.

Thomas, Nick. 1980. Point Conic Space and Path Curves. *Mathematical-Physical Correspondence*. 31.

— 1984. Das Zusammenwirken von Raum und Gegenraum zur Erzeugung von Wegkurven [Working of space and counterspace to produce path curves]. *Mathematisch-Physikalische Korrespondenz. 133.*

— 1999. *Science between Space and Counterspace.* London. Temple Lodge Press.

Tompkins, Peter, and Bird, Christopher. 1989. *Secrets of the Soil.* London. Harper Collins.

Van Mansfeld, J.D. 1986. *Virbela Waterprojekt 1982.* Netherlands. Wageningen University.

Wagenaar, Walter. 1985. Beweging: Sturing in levensprocessen van het water. Ondersoeksverslag 1983–84 [Movement: influences of water on life processes]. *Wirbela Waterprojekt.* Netherlands. Warmonderhof.

Watts, Alan. 1975. Tao, The Watercourse Way. London. Arkana.

Wilkes, A. John. 1993. Flow design research relating to Flowforms. in *Chaos, Rhythm and Flow in Nature, The Golden Blade.* Edinburgh. Floris Books.

—. 2001. Water as Mediator. *Elemente der Naturwissenschaft.* No. 74 Dornach.*

Will, Reinhold D. 1993. *Geheimnis Wasser.* Munich. Knaur

Zoeteman, Kees. 1991. *Gaia Sophia.* Edinburgh. Floris Books.

Index

Entries in *italic* indicate pictures

Ton Alberts Garden *157*
tree *36f*
Trevelyan, George 9
Trousdell, Iain 158, 161, 158, 187, 196, 200
Trussocks Hotel 110
Tube Flowform 75
turbulent water movement 22, 34, 36, 47, 49, 54f, 76, 88f, 208

Uhlvund, Jürgen, 13, 226
Ulm *157*
Underwood, Leon 9

veils 47
Vestergaard, Ole 168
Vidaråsen 106
Virbela Atelje 110
Virbela Rhythm Research Institute 170, 173

Virbela Screw 145, 147, 171
viscosity 50, 92, 134, 144, 208, 229
vitality 89f, 166
vortex 29, 48, 55, 145, 149, 173
Vortex Flowform 69, 106, *111, 187, 190, 204*
Vortex Four Cavity Flowform 191
vortex ring 47
vortex suction pump *149f*, 171f
vortices 51
—, path of *(Wirbelstrasse)* 42, 49f, 181

Warmonderhof project 106, 135–141
water cycle 25, 26f, 41, 88f, 140, 172
Water Garden, Chengdu 112
water as mediator 15f, 35f, 79, 82, 89, 93, 170, 173, 218, 223, 236
water sacrifice 35, 172
water in science and art 224
water transport 166

waterfall 26, 45, 48, 61, 65, 66, 71f, 74, 88, 115, 118f, *126f,* 140, *154, 157,* 162, 188, 200, 202
Weidmann, Nick 12, 114, 122, 130, 162, 201f, 204f
Weleda, Arlesheim *157*
Wellhead Stackable Flowform *162*
Wellington, N.Z. *158*
Wells, Nigel 11f, 99, 112f, 119f, *121f,* 129, 135, 155, *159,* 187, 194f
Wheel Flow Unit 147
Whicher, Olive 59, 173
Wilkes, Thomas 13, 225
*Wirbelstrasse (*path of vortices) 42, 49f, 181
Withington 112

Zijpendal, Park 114

Floris Books

For news on all our **latest books**,
and to receive **exclusive discounts**,
join our mailing list at:

florisbooks.co.uk

Plus subscribers get a FREE book
with every online order!

We will never pass your details to anyone else.